PARTICIPATORY ACTION RESEARCH IN NATURAL RESOURCE MANAGEMENT

D0169153

PARTICIPATORY ACTION RESEARCH IN NATURAL RESOURCE MANAGEMENT

A Critique of the Method Based on Five Years' Experience in the Transamazônica Region of Brazil

Christian Castellanet

Groupe de Recherche et d'Echanges Technologiques, Paris, France

Carl F. Jordan

Institute of Ecology, University of Georgia, Athens, Georgia, USA

USA	Publishing Office:	TAYLOR & FRANCIS *A member of the Taylor & Francis Group* 29 West 35th Street New York, NY 10001 Tel: (212) 216-7800 Fax: (212) 564-7854
	Distribution Center:	TAYLOR & FRANCIS *A member of the Taylor & Francis Group* 7625 Empire Drive Florence, KY 41042 Tel: 1-800-624-7064 Fax: 1-800-248-4724
UK		TAYLOR & FRANCIS *A member of the Taylor & Francis Group* 27 Church Road Hove E. Sussex, BN3 2FA Tel.: +44 (0) 1273 207411 Fax: +44 (0) 1273 205612

PARTICIPATORY ACTION RESEARCH IN NATURAL RESOURCE MANAGEMENT: A Critique of the Method Based on Five Years' Experience in the Transamazônica Region of Brazil

Copyright © 2002: All rights reserved. Printed in the United States of America. Except as permitted under the United States Copyright Act of 1976, no part of this publication may be reproduced or distributed in any form or by any means, or stored in a database or retrieval system, without prior written permission of the publisher.

1 2 3 4 5 6 7 8 9 0

Printed by Sheridan Books, Ann Arbor, Michigan, 2002.
Cover design by Ellen Seguin.

A CIP catalog record for this book is available from the British Library.
♾ The paper in this publication meets the requirements of the ANSI Standard Z39.48-1984 (Permanence of Paper).

Library of Congress Cataloging-in-Publication Data

Available from the publisher.

ISBN 1-56032-979-3

CONTENTS

PART III
LESSONS FROM THE PARTICIPATORY ACTION RESEARCH IN THE TRANSAMAZÔNICA

APPENDICES

PREFACE

In the 1980s and early 1990s, I had the opportunity to observe a number of development projects in the Amazon region of Brazil. Some of the projects were designed to improve the life of colonists in the region, while others were focused on management of natural resources. All the projects were "top down" in the sense that project design and direction were carried out by high-level corporate or government sponsors at headquarters far removed from the affected areas. There was little or no input from people who actually lived in the area, people who had first-hand knowledge of the social and environmental problems in the areas to be developed. As a result, most of the projects were not as successful as they might have been.

During the course of my visits, I became aware of the Programa Agro-Ecologico da Transamazônica (PAET). This effort was sponsored by the European Community and Groupe de Recherches et d'Echanges Technologiques (GRET), a French nongovernment organization. The objective of PAET was to improve farming practices and management of natural resources along the Transamazonian Highway (really just a dirt road) near Altamira, Brazil. The focus was on the community-based participatory action research (PAR) approach to development.

I was interested in whether PAR might be a better method than the approach used by other projects that I had studied. I met the GRET project leader, Christian Castellanet, and persuaded him to take a PhD with me at the University of Georgia with the condition that, for his dissertation, he would analyze and report on the strengths and weaknesses of the PAR approach to development based on the Altamira project. He agreed. This book is based on project documents, transcripts of meetings, interviews, and personal notes that Christian took during his five years in Altamira.

The project had some successes and some failures and, as is true for all projects, many aspects were unique to the place and time. However, insights regarding strengths and weaknesses of PAR may have a more universal applicability. The interactions that developed and the problems that arose between the research team, the local community, and the focus group (in this case, the farmers' organization) may be similar for any PAR concerned with resource management. So that future projects using PAR might derive the most benefit from the Altamira experience, we focus on analysis and discussion of the method, with the project itself as a backdrop against which PAR is used and evaluated.

Carl F. Jordan

INTRODUCTION

The involvement of scientists in public affairs has grown since the end of World War II. Initially, it was restricted to questions pertaining to military capabilities (Rotblat 1982), but this involvement later included other areas such as development, human rights, demography, and environment. Concern about the latter increased sharply after the Club of Rome report, in which stark consequences were predicted if the world's population exceeded the environmental limits of growth (Meadows et al. 1972).

As technology has continued to progress and apparently insoluble social problems have continued to develop, educated citizens have begun to realize that science has given us a formidable capacity to manipulate the physical world, but a very low capacity to intervene in social problems such as the growing gap between the rich and poor, unemployment, population control, growing violence, and social instability (Lakoff 1980). The development of human wisdom and the capacity to better organize and cooperate has not paralleled the development of science. Science has shown us that the more we manipulate things, the more we run the risk of destroying our own habitat, or at least damaging it so much as to make our existence miserable. As Rabelais put it 400 years ago, "Science without Conscience is but the Ruin of the Soul."

Both the perception of science by the public and the perception that scientists have of their role in society are changing. Although we still depend on science and technology for the operation and improvement of our material culture, few still believe that science has the answer to all human problems. Indeed, we are now confronted with a set of problems that are increasing in number and intensity. Many are the result of technological and industrial developments. Science, although a necessary element of their solution, will not be sufficient for *their* solution. After World War II, one could imagine science advancing boldly, steadily rolling back the frontier between knowledge and ignorance. Now we must cope with our ignorance of the ramified effects of science (Ravetz 1989).

Complex social and environmental problems are not amenable to the usual reductionist/disciplinary methods of science. Scientists trying to solve such problems have to integrate uncertainty into their conclusions (Jordan and Miller 1996). Ravetz (1989) and Roqueplo (1996) showed how the uncertainty associated with most complex environmental problems affects scientists' behavior and places them in an uncomfortable position as "experts." Scientists engaged in environmental issues also try (although they are not always conscious of it) to intervene in order to change some or all of society. More precisely, if one adheres to an individualist perspective, he or she tries to change other people's behavior. Many scientists do not have sufficient training and background in social science to coldly and objectively analyze their own position and concepts in this context (Bailey 1996).

Participatory action research (PAR) is a method that has been proposed to overcome the problems inherent in traditional scientific approaches to problems of development and resource conservation. However, PAR has not been tested adequately in the context of natural resource management; it is not yet clear if it presents a viable alternative to the traditional approaches. The work on which this book is based has presented an opportunity for an in-depth examination of the method. The results will be of interest to scientists and policymakers who are trying to increase the efficacy of programs intended to solve environmental problems. The lessons learned may help them achieve their goals.

PART

I

BACKGROUND

Approaches to Resource Conservation

Traditional Scientific Approaches

Natural scientists who participate in practical measures to solve environmental problems generally take one of two contrasting approaches: the moralist/educational view or the authoritarian view. The moralist/educational view assumes that human beings are willing and able to change their values and subsequent behavior once they understand the long-term consequences of their actions on themselves and others (Leopold 1949, Orr 1992). The authoritarian view holds that politicians, who are supposed to be able to guide the rest of society, should listen to enlightened scientists who can tell them of the best policies (e.g., Myers 1979, Wilson 1992). This view is in the tradition of Auguste Comte (1854), who suggested that scientists should be in charge of government. Both approaches can be considered "top down," that is, a blueprint for local situations.

The Moralist/Educational Approach

Those who take the moralist/educational philosophical line usually choose to work in education, mass communication, or public relations. The long-term impact and efficiency of this type of effort is difficult to evaluate.

On the one hand, it is naïve to believe that the basic values and attitudes of a culture change drastically in one or two generations. Historians of culture note that cultural traits generally change slowly, more slowly than technology and the environment in which the technology emerges. On the other hand, new religions and political revolutions have resulted in drastic changes in ethics. Public campaigns can also result in a change of values. The growing consumer interest in "green" products demonstrates the latter point. However, the recent debate about US oil consumption and the rejection of any oil taxation show that the common good remains marginal compared with individuals' values. It seems unlikely that the "greening" of citizens' opinions is sufficient to profoundly change the type of development that has prevailed over the last centuries.

Another limitation of the moralist position arises from cross-cultural dialogue, that is, intervention in foreign countries with different cultures. Often, legitimate concern about education in international cooperation programs translates into ill-conceived and inefficient schemes of "environmental education." These efforts frequently have, as a basic tenet, a naïve view of education. Unwise use of natural resources, it is believed, is the result of people's ignorance of the functions and values of nature. Those who hold this belief are naïve in various senses: (1) by believing that local people don't know the value of natural ecosystems and how they can benefit from them; (2) by forgetting that natural ecosystems also pose a threat (poisonous snakes, disease-carrying insects); and (3) by not understanding that immediate survival may depend on exploiting natural resources without regard to sustainability. The small farmer who burns his forest to replace it with pasture is not fundamentally different from the ecologist who uses a big car to go to his or her laboratory. Both know that they are using natural resources in an unsustainable way, but the farmer may have an idea about how he will develop a new agricultural system after the forest is gone. The ecologist should know that there is no known way to reverse the build-up of carbon dioxide.

Informing people about the consequences of their actions is not totally useless. For example, a good information program can promote awareness that certain resources are apparently limited. Thus, people may come to accept and even support new rules or policies that will restrain their use of these resources. However, changing actions as a result of an information program is quite different from changing fundamental behavior as a result of cultural evolution.

The Authoritarian Approach

In the authoritarian approach to development, political, economic, or bureaucratic authorities, based on recommendations by consultants, often decide on a project. In natural resource management, the consultants are usually scientists. However, scientists are often divided, both on the practical measures that should be taken to solve a particular problem and on the exact nature and gravity of the problem.

To solve scientific uncertainty in natural resource management, a proposal is often made to establish huge interdisciplinary research programs to reach clear conclusions on the nature of the problems, and solutions are recommended. The results have generally been weak from a scientific point of view and even more disappointing when it comes to practical decisions and results. The problem of interdisciplinary research has been discussed by various authors, particularly Pivot and Perocheau (1994), Godard (1992), and Rhoades (1984).

Another problem with authoritarian solutions is that politicians' capacity to influence society may be much less than is commonly believed. It is difficult to enforce a law or rule that is not accepted by the majority of citizens, even in the most dictatorial regimes. In the case of protection for national reserves and parks in developing regions, Sayer (1991) concluded that legal protection is seldom sufficient to permanently guarantee the integrity of protected areas. The local population frequently sees parks as a restriction on its traditional rights being imposed by a distant, central government. When this happens, protected areas lose popular support and their condition quickly deteriorates.

An example of the authoritarian approach is given in McKinnon et al. (1990) in *Management and Guidelines for Tropical Protected Areas*. Most of this manual discusses the planning and establishment of protected areas exclusively on the basis of discussions among scientists, nongovernment organizations (NGOs), and governments. This book notes that park authorities should cooperate with local populations in finding ways to obtain some economic returns from the protected area. However, it doesn't point out that local populations and authorities can negotiate issues such as boundaries of a protected area and rules for management of the reserve. Local people are to be invited only to "cooperate" in project implementation, not to participate in the project design. As a result, conflicts are common and the resource management plan hardly ever survives (Sayer 1991).

Difficulties in the authoritarian approach arise partly because most of the staff of conservation projects and organizations consists of conservation biologists, foresters, and wildlife managers. They tend to separate the human component of conservation projects from the biological component, to which they give more attention and priority. They fail to recognize that, although the ultimate goals of conservation efforts may be driven primarily by biological theory and ecological research, the process by which conservation is achieved is overwhelmingly social and political (Bailey 1996). In our opinion, neither the educational approach nor the authoritarian approach can be effective in solving the world's environmental problems. Other methods must be used, based on the participation of all stakeholders and on negotiation and compromises among these different actors (participants) and the state.

Conventional Methods of Intervention for Natural Resource Management

Most environmental problems are the result of inadequate management of natural resources at the local level. Various types of intervention can be proposed to improve local management, with a view toward broader and more long-term interests. The methods of intervention proposed to reach this objective can be diverse, and their respective merits and weaknesses are still in debate. They range from highly publicized demonstration projects, organized at the local level and visited by public officials, to programs linking environmental public agencies with mass environmental education. A particular case is the establishment of "nature reserves" or parks, for which either an authoritarian or a participative approach can be used.

Nature Reserves and Buffer Zones

Despite the growing use of participative rhetoric in the discourse of international conservation organizations such as the International Union for Conservation of Nature and the World Wide Fund for Nature, in practical terms local participation is generally restricted to discussion about the type of compensations the local population might receive from the park or reserve authority in return for their losses, which may include restrictions in access to reserve land and natural resources (see Taylor and Johansson [1997] on the Masai participation in Ngorongoro protected area).

In a study organized by a conservationist organization, Hannah (1992) concluded that most conservationists believe that local populations should participate in management decisions concerning African parks and should share in profits from tourism. This participation and sharing are necessary for the long-term viability of protected areas. However, these are still ideas; few parks in Africa really allow local populations to participate in decision-making.

One of the main difficulties is deciding on the most appropriate political structures to represent the interests of local populations. Support of local organizations sometimes results in conflict with the national "elite," who derive some of their profits from exploitation of local manpower and natural resources. For example, in the Dzangha Reserve in the Central African

Republic, efforts to transfer some of the tourism income to the local Aka Pygmy's groups met fierce opposition from local political leaders and public servants, who derive various benefits from logging in the reserve and from illegal trade of furs and wild animals (Caroll 1992, Colchester 1995).

There is no simple method of implementing the conflict management strategies that part of the conservationist community now feels to be necessary (Kemf 1993). However, discussion with local communities regarding how they can derive direct benefits from the park (either from tourism or by better exploiting part of its natural resources in a controlled, sustainable way) certainly indicates progress toward acceptance of parks by local people (Ledec and Goodland 1989). Nevertheless, it is not a guarantee of success.

One of the first and oldest efforts to effectively negotiate environmental management with local indigenous populations appears to have been conducted in British Columbia. Conclusions from these efforts focused on the necessity of training local people to do the following: to participate as equal partners in a management team; to make decisions based on consensus and not on a simple majority; to avoid later debates and clashes during local elections; to formalize agreements with unambiguous written contracts; to establish mechanisms that increase community income and participation; and, finally, to establish common research programs (Davey 1993). Another important lesson is that the local population should be associated as early in the process as possible to avoid unnecessary misunderstandings and conflicts. Whenever possible, the local population should be guaranteed its territorial rights. Furthermore, within their territory, they should be authorized to use its natural resources as long as such use does not destroy the resource.

With these conclusions, the conservationists adopted a perspective that reconciled with those who supported indigenous peoples' rights in the management of common goods by local communities (as discussed by Ostrom 1990).

We must avoid, however, idealizing the indigenous culture and its supposed "harmony with nature." Numerous examples show that indigenous peoples are quite able to destroy their natural resources after contact with the market economy and dominating cultures and having access to modern technologies. Anthropologists have not yet found any conclusive link between the indigenous religion and management of natural resources. For example, there is no conscious effort to conserve natural resources in Amazonian Indian societies. As a result of their traditional political and subsistence system, which encouraged permanent mobility of small Indian groups, they maintained a low pressure on natural resources. However, when these Indians become sedentary and gain access to new technologies, they may quickly exploit the local environment (Colchester 1995).

Demonstration Projects and Technology Transfers

Demonstration projects are commonly included in the authoritarian and educational approaches. The basic idea is to encourage local initiatives that lead to a more sustainable use of resources, whether in agriculture, forestry, or fisheries. "Demonstration" is meant to illustrate the superiority of these projects and it is assumed that they will spread by virtue of their example. The projects suffer, however, from the incorrect assumption that just because certain technologies exist, they can be successfully applied in the field. This model has been outdated since the 1960s. Agronomists and anthropologists who have studied diffusion of new agricultural methods have clearly demonstrated that, in most cases, proposed technologies are not adopted because they simply do not meet the needs and requirements of potential users. The problem is not how to get the technology to the small farmers, but how to design or find technologies that are of some use to them (Shaner et al. 1982, Roling 1988).

There is no reason to believe that the diffusion of technologies in the field of forestry, fishing, or extractivism would be much different from the diffusion of agricultural innovations. To be efficient, a project aimed at the diffusion of new technologies should do the following:

- Link research, extension, and social experimentation
- Support social experiments in a progressive and continued fashion (start small, evaluate the results regularly, and grow steadily if the results are encouraging)
- Avoid subsidizing material, equipment, or production itself, since it completely distorts the economic sustainability of the technology that is introduced
- Be flexible in scope and nature, but with regular outside evaluations

This is more or less the opposite of what is done in most demonstration projects, which use the blueprint approach to project planning. This approach usually has limited duration but high external funding and does not leave any opportunity for participatory experimentation and learning.

Research on Agricultural and Agroforestry Systems

By applying known scientific principles, agronomists and ecologists often design agricultural systems that are environmentally superior to systems in use—with regard to nutrient recycling, erosion control, pest reduction, and sustainable production. The approach usually is to design and test the proposed system in a controlled environment such as a research station. Once its efficiency is proven, the system is transferred to local farmers. However, the farmers for whom they are intended do not usually adopt such systems on any significant scale. A famous case is that of the International Institute for Tropical Agriculture in Nigeria, where various alley cropping systems have been tested for years but local farmers were never convinced to adopt them (Lal 1991). The International Center for Research in Agroforestry has been criticized for its lack of success in getting farmers to adopt new agroforestry systems. However, traditional agronomists have not done any better. For example, the high-input approach of the research station in Yurimaguas, Peru has also been unsuccessful (Sanchez et al. 1987).

It is not through ignorance, tradition, or passivity that farmers do not adopt the proposed systems. These systems might perform nicely from a technical point of view, but they usually do not fit into the farmers' constraints. They are either too labor-intensive or too risky, or they need a long-term investment that the farmers cannot afford (Fujisaka 1991). Floquet and Mongbo (1994) observed in South Benin, in West Africa, that the same farmers who do not adopt the improved technologies proposed by researchers are actively experimenting with their own indigenous innovations in methods of soil tillage. The conclusions are that: (1) researchers should make an effort to identify the existing local innovations and build on them rather than ignore them and (2) improved systems should be designed in an incremental way and in cooperation with the farmers from the start, rather than at the experimental station.

An example of promising alternatives such as agroforestry systems in Amazonia and their analysis and diffusion to other groups of farmers is given by Anderson (1990a). He recognizes that, although the technical and ecological aspects of these improved systems are relatively well known, the economical and social conditions under which they become feasible are understood much less. It is relatively easy, however, to pinpoint the main limitations that restrict the diffusion of the alternative systems described in this book. For example, in the agroforestry system (Anderson 1990b), the critical factor is access to a very specific urban market (in Belém) for fresh açai palm. It would not be possible to implement such an agroforestry system more than 20 miles from Belém. In Japanese agroforestry methods at Tome Açu, described by Subler and Uhl (1990), high capital and technical know-how are needed, but access to Belém markets for poultry, fresh fruits, etc., is also a factor. The Japanese-Brazilian community controls some of these markets, so they would probably not be open to other farmers. The alternative forestry methods proposed by Harsthorn (1990) have not been tested long enough to draw any conclusion about their economic viability.

In summation, the main limitations in the diffusion of improved systems (in terms of ecological sustainability) in Amazonia is not the need for new research into their technical and ecological aspects, but the socioeconomic constraints such as access to market and transportation costs, land market values, capital, training for farmers, and farmers' organizations (Sawyer 1990). This situation does not necessarily imply that the farmers already know the possible alternatives. Exchange visits and training can certainly be beneficial, especially in a large region such as Amazonia with highly dispersed settlements of various origins.

The Pilot Program for Amazonia

In 1991, a pilot program to conserve Brazilian rainforests was approved by representatives of the G7 countries (Batmanian 1994). They agreed that assistance of about $280 million (US) would be provided in the form of grants, technical cooperation, and loans. After this initial commitment, the executive directors of the World Bank established the Rain Forest Trust Fund and assigned several staff members to coordinate the pilot program and the Rain Forest Trust. The main initiatives of the program were to:

- Establish information systems and training activities to prepare a zoning plan for the Amazon region
- Provide management plans for Conservation Units, such as indigenous reserves and extractive reserves
- Provide support to the Emilio Goeldi Museum and the National Institute of Amazon Studies (INPA) for them to be centers of excellence for scientific research
- Establish monitoring and enforcement of existing environmental laws
- Encourage the rational use of natural resources
- Support environmental education
- Develop demonstration projects

Many of the proposed lines were conceived on a top-down basis and are basically conducted by federal government structures. Therefore, the program is an interesting mixture of locally based demonstration efforts or pilot activities and research, combined with government capacity to monitor and control the processes.

Scientist participation was to be mostly in the form of expertise, in which scientists are consulted regarding which projects are the best to support within each subprogram line. Support for the regional research centers seems to stem from the premise that academic research will help in better understanding what is happening in Amazonia and, therefore, in better directing efforts to save some of its natural richness. However, there is no provision for linking basic research to project activities or to local demands and proposals.

Demonstration projects in principle encourage the participation of local populations to the elaboration and execution of alternative methods of natural resource management. Most projects are relatively small and are presented by NGOs or local organizations. However, the project selection and funding process do not permit effective participation of the concerned populations. These projects have a rigid framework of three years, the use of the funding has to be rigidly planned in advance, and there is no provision for a preparation or pilot phase. There also is no evaluation of the real participation of the populations in the project planning. The lack of in-depth regular evaluations of the field projects also limits their heuristic value. As a result, many initiatives are likely to remain "small white elephants."

Unfortunately, the sectorial approach that characterizes the pilot program (with a specific advisory committee for each line, separating research institutions from NGOs and government representatives from project managers) does not favor this exchange of information and debate.

Participatory Approaches

In recent years, new approaches give consideration to the perspective of the local people, sometimes called "bottom up" or "participatory." In contrast to the authoritarian approach, which is more common among natural scientists, the participatory approach is more common among social scientists. This approach considers that the scientist is also part of the society that he or she hopes to change and that he or she should contribute to the debate on an *equal level* with local populations and other citizens, including professional politicians. The necessity of establishing real communication through a two-way dialogue comes from several premises:

- The complexity of development problems is so great that no single specialist can pretend to know and understand all relevant aspects of the problem.
- It is recognized that popular knowledge and local professional but non-scientific knowledge have value.
- Most environmental problems cannot be resolved solely through either top-down (authoritarian) or bottom-up (participatory) approaches. Top down is the planning process in which technologies, projects, or policy proposals are devised by experts or politicians and then transferred or imposed on local populations. In contrast, bottom up is the process in

which local needs and proposals are the basis on which decisions are taken (Rhoades and Booth 1982, Chambers et al. 1989).

The first two of the latter premises are easily demonstrable and widely recognized among social scientists, but the third premise is more subjective and may be considered a philosophical option rather than a demonstrable rule. However, one can argue that if the first two are true, the third should be true also. In addition, it seems obvious that cooperation of the local population is essential for a good diagnosis of the environmental problem to be treated, but that a higher authority is often needed to give legitimacy to any proposed solution. In addition, solutions have to seem reasonably legitimate to the concerned population to have some chance of success. Therefore, the participation of all concerned groups in the identification and resolution of the environmental problems is essential.

Rapid Participatory Environmental Appraisal

Some international NGOs such as the International Institute for Environment and Development (London) initially specialized in approaches to rural development and developed a method called rapid participatory rural appraisal (often abbreviated simply RRA) which, in principle, guarantees population participation during the initial planning of a given project, spelling out its own priorities and suggestions on "how the project should go." This approach was in fact an adaptation of a method developed earlier by International Agricultural Research Centers called "Sundeo." This method was adapted to natural resources management recently under the name of "participatory environmental appraisal." Such methods bring together, in a relatively short period (15 days), all information and demands expressed by the community. Usually, the work is carried out in groups, sometimes with the separation of special interest groups such as women and young people, and with the participation of local technicians and government agents. The participatory environmental appraisal requires the intervention of a qualified facilitator. Obviously, it represents an advance in relation to the practice of project planning by experts and government officials without any popular participation except to answer questions. However, the participatory resource appraisal has its limitations and criticisms have been severe, mostly due to the lack of caution by its advocates, who tend to present it as a miracle solution (Fall and Lericollais 1992, Cornwall 1992, Olivier De Sardan 1995). The main problem is that these methods are supposed to help local people to express their demands autonomously, whereas in fact the facilitator always influences this formulation—often unconsciously. The simple fact that a team comes to a village to organize an appraisal creates expectations, based on the past experience of the villagers with other project managers, colonial authorities, or government agencies. Therefore the demand is shaped by this perception.

Other difficulties have to do with the fact that too often the RRA has been based on an idyllic vision of the village "community." Experience shows that more often than not, there are internal community differences and conflicts. The control that the local elite maintains over the rest of the local people often results in a false unanimity. Most practitioners now agree that the rapid appraisal might be an interesting method as a first contact but cannot replace professional ethnographic work in analyzing the local power relationship and the various strategies of different groups, including strategies for dealing with outsiders, researchers, and project managers.

The community itself is never separated from the rest of the society; therefore its dynamics also depend on other communities and other public agencies that may have some claim on the natural resources and that should also be heard in any planning exercise. In short, conflicts do exist and are the rule rather than the exception in terms of natural resource management. If conflicts exist, a rapid exercise, whatever its content, can hardly be expected to solve them through a new consensus. In terms of effective empowerment of the local people, it is also obvious that a long-term perspective is needed and that a rapid participation will not change fundamental power relationships (Wright and Nelson 1995).

In a recent review of participation in planning exercises organized for National Conservation Strategies and National Sustainable Development Strategies, Bass et al. (1995) concluded that participation of all categories of stakeholders was a condition to greater success of these strategies. However, the study lacks precision in the way this success has been evaluated. A point that stands out is that they recognize that the number of "win-win" possibilities in sustainable development (i.e., situations in which all different groups can gain from the new proposed plan) is limited and that conflict resolution is invariably required. This means that the solutions to these conflicts have been imposed by the governments and, therefore, risk of further claim by the losers is possible.

Conflict Management and Mediation Theory

Conflicts about environmental matters often involve at least three parties: the local citizens, numerous but poorly organized; a big company with an industrial or commercial project; and the local and national authorities. Susskind and Cruikshank (1987) describe how a mediator can help to resolve such conflicts through negotiation. The mediation approach draws partly on game theory in which opposing parties assess their potential gains and losses for each choice that can be made. In the case of a lawsuit, the choices might be to go to court, continue the negotiation, or accept a proposed agreement. An agreement should be made when it becomes the most acceptable option for all concerned. The mediator's job is to help the various parties in making this assessment and in encouraging the mutual understanding of each other's priorities and limits so that new innovative proposals can be made to reduce the losses and increase the gains for all parties. Mediators also have to make special efforts to help organize the poorly represented groups, nominate their representatives for negotiation, and verify that these representatives maintain close contact and discussion with the rest of their group.

According to Maser (1996), there are two basic approaches to mediation (or facilitation) in environmental conflicts. One approach concentrates on problem-solving, the other on the moral transformation of the disputants. The problem-solving approach often uses the stakeholder analysis method.

Multiple Stakeholders' Analysis

When resources are used by various social actors, either directly or indirectly, it is necessary to identify the groups concerned, their specific interests, what value they give to the resource, and their areas of conflict or cooperation. This is the scope of multiple stakeholders' analysis, a method initially developed by environmental economists with the objective of evaluating the different values of natural resources to different groups of users (Freeman 1984).

Stakeholder analysis in the field of natural resources uses the following approach (Grimble and Wellard 1997):

- List the present and potential stakeholders at the different institutional levels—local, regional, national, and international. For each stakeholder, list what interest each one has in the resource ("issue of environmental interest").
- Classify the stakeholders in terms of importance (based on policy criteria) and influence (their power to influence an outside intervention). For example, small farmers may be most important to a program that wishes to benefit the majority of the population, but they have limited influence locally; big landowners have a limited importance, but great influence.
- Identify the existing conflicts or cooperation among stakeholders and the trade-off of possible decisions made by policy-makers or local groups.

The main interest of stakeholder analysis is to highlight the possible consequences and reactions of the stakeholders to any decision affecting natural resources. It is therefore essentially a method designed to assist experts, project managers, and policy-makers in assessing the efficiency, feasibility, and effectiveness of different possible decisions in terms of project selection or policies.

Multiple stakeholders' analysis has often been tried in developed countries (Bacow and Wheeler 1984). However, there have also been experiences with resource management in developing regions. Warner et al. (1996) described such an experience conducted in game management areas in Zambia and Pimbert et al. (1996) described another analysis conducted around protected wetland in India and Pakistan. Stakeholders included local villagers and conservation authorities. In both cases, Rapid Rural Appraisals (RRAs) helped to identify the local villagers' knowledge, views, and concerns about the protected areas, as well as their proposals for solving some conflicts with conservation authorities. Some of the proposals appeared to meet both parties' interests and were able to be implemented quickly. Others could be used as a starting point for wider negotiations on policy.

The Platform Approach

The platform approach is a strategy that aims at transforming participants' values and perceptions of the problem so that an acceptable solution for all parties becomes more likely. Of special importance are the understanding and recognition of other disputants' attitudes, values, and vision of the problem. This permits them to shift from an attitude of destructive conflict to one of constructive negotiation.

Roling (1994) proposed to include "knowledge systems approaches" within conflict management methods, which are developed by the communication researchers to create "human platforms to manage natural resources" in the rural environment. What this means is that special attention has to be given to conflicts between the local farmers' knowledge and view of the ecosystem to be managed and the technical-scientific vision. As a result of this dialogue, a richer picture is likely to emerge and the various stakeholders can also progressively build a joint perspective on the natural resources at stake. Reaching an agreement through this effort about the facts observed and mechanisms involved, such as changes in the water cycle as a result of deforestation, facilitates further negotiation.

This platform approach, which can be characterized as "cognitive constructivism," is grounded on Habermas' (1984) theory of "communicative action." It states that societal change (through consensus building) can result from rational communication among various actors who agree to reach a noncoerced mutual understanding. To achieve rational communication, participants must agree to allow any and all statements to be questioned and to work toward resolving the validity of dubious statements through open debate. Debate can be based on personal knowledge, expert testimony, published expertise, or experiences shared by discourse participants (Webler 1995). Of special importance is the discussion of long-term goals and strategies aimed at finding a common long-term goal that will be acceptable to all participants.

This cognitive constructivist approach is used today by various researchers, who describe it as "soft system methodology" (Checkland and Scholes 1990), "systems learning" (Bawden 1991), or "participative ecodesign" (Ison et al. 1997). However, relatively few applications have been made in the environmental field so far and most are in developed countries, such as sustainable agriculture in the Netherlands (Roling 1994) and the Landcare Movement in Australia (Campbell 1994). William et al. (1994) describe their approach as a "learning process," which has been used successfully in solving local environmental problems such as river-basin management and co-management of ranching and wildlife in Oregon. However, experiences in developing countries are limited and extremely recent (Grimble and Wellard 1997, Bass et al. 1995).

The Patrimonial Approach and Common Property Management

The question of the sustainability of natural resources owned or controlled, in common, by a certain group of users is a particular case of resource management that has been studied extensively by environmental economists, anthropologists, and lawyers. Contradicting Hardin's (1968) famous statement that all commons are bound to be tragically overexploited, a body of literature has shown that one should not confound free access with common property, and that many examples of efficient and sustainable forms of management of common properties can be found around the world—in traditional communities as well as in industrialized countries (McCay and Acheson 1990). However, situations in which common properties are not managed satisfactorily from both an ecological and economical point of view are also common.

Ostrom et al. (1994) have studied the conditions for the emergence and sustainability of what Ostrom calls "institutions for collective action," or common property management. This includes the capacity of a given group (either formal or informal) to design and adopt a set of rules (either written or oral) and control their application. The researchers concluded that a set of economical, cultural, and institutional conditions strongly influence the success of these collective institutions. Among the most important are the following:

1. *Economic aspect* (based on the game theory): The cost of enforcing the rules must be much lower than the benefits of the common management and the potential cost of cheating must be higher than the benefit of the cheating. This, in turn, depends partly on the nature of the common resource. For example, it is easier for the community to control access to a nearby pasture than to game in the forest.
2. *Social aspects*: Groups have more chances to elaborate and accept common rules when they share a common culture and history than when they are multicultural. Small groups (nested organizations) are more successful than big ones.
3. *Political aspect*: Common agreements are much stronger when they are supported by the government or the judiciary system.

Culture influences whether natural resources are considered common property. Cree Indians believe that it is arrogant to try to manipulate animal populations. They do not consider game animals a common property but a gift of the gods to men. As a result, they reject management measures (Berkes 1990). Even in cultural situations more similar to ours, the concept of a common good as universal is not obvious. Olivier De Sardan (1996b) argues that there is no "communal" tradition of natural resource management in West Africa; other types of traditions can be observed there. One is based on the patriarchal system, another is an aristocratic tradition, a third is the result of modern state impositions, and the fourth tradition, also modern, is the clientelist tradition, often linked to political parties. Conflicts between traditions probably constitute one of the main reasons for the disappointing results of large-scale efforts undertaken in the 1990s to promote local governance through the gestion de terroirs (communal land management) approach.

Culture also influences whether common resources are consciously managed. Indigenous peoples are often assumed to be cultural conservationists. However, they do not have the same definition of nature that we do. It is possible that some traditional hunter peoples do, in the long term, manage their sustainable resources, but this is not necessarily a result of their respect for animals or wise vision of the future. It may be an unwilling and probably unconscious result of combined social and religious rules about when a group should change its location, when it should divide, and about when aggressive behavior toward other groups is appropriate, as Colchester (1995) observed in the Venezuelan Yanomami. Algonquian Indians, who thought that game animals killed by hunters spontaneously regenerated after death, probably encouraged indiscriminate killing before any European influence occurred (Brightman 1990).

The patrimonial approach holds that proper management of natural resources requires that these resources should not only be identified as common goods, but should also be seen as part of a heritage that the group needs to transmit to its children (Ollagnon 1989). How this consciousness can be encouraged or increased in a practical way is an open question, although some practitioners hope that participatory approaches supported by the state can help in this direction. Although projects with this objective have started in Madagascar (Bertrand and Weber 1995), Olivier De Sardan (1996b) is skeptical about the chances for success in Africa. He feels that these processes will take a long time and are linked with a new concept of the state government and its relation to the citizens.

Limitations of Participatory Approaches

Participation is not the miracle solution to the problem of reconciling development and environmental priorities. Participation is "a very warmly persuasive word," but also a very deceitful one that is used to describe very different sets of relations (Nelson and Wright 1995). Participation may mean simply the *involvement* of local people in the execution of a given project. Or, at the other extreme, it may mean that these people not only decide what the project should be about, but directly manage the resources obtained from the aid donors. It could also mean that the local people have to build up their own capacities to promote "self-organization development" independently from the state or outside assistance (Rahman 1993).

Since most environmental problems involve some conflicts between various types of stakeholders and generally at various hierarchical levels (state, county, municipality, village, farm), purely bottom-up participatory approaches are not usually able to solve the problems. Some intervention of higher-level public authority is necessary to solve the conflicts or at least to enforce the agreement to which the stakeholders may reach after negotiation. We therefore advocate participatory action research (PAR) because it is an intermediate between the top-down and bottom-up attitudes. It aims at both contributing to changes in individual views and values and contributing to new policies and laws. PAR is an option often advocated among political scientists and development agents today. However, strong advocates of other approaches remain. For example, some high-level civil servants consider that economic laws dominate the world and that ground-level development projects have a very limited impact on societal change. On the other hand, some sociologists and educators consider that all important social and political change comes from a person's change of attitude during everyday life.

Evaluation of an Alternative

Methods of natural resource management are needed, which are able to successfully treat complex problems involving high risks, uncertainties, and social dynamics and to promote the participation of all actors involved. A crucial question is how to bridge the gap between the growing amount of scientific and technical knowledge and our incapacity to solve some of the seemingly most simple environmental problems. A corollary problem is how concerned scientists can become involved in resolving these problems. Various methods and approaches have been proposed in recent years to remedy problems inherent in other approaches to development of conservation projects, but they are still at a design or testing stage. The objective of this book is to discuss the validity of one of these new proposed approaches, participatory action research in the field of natural resource management. It differs from previous types of participatory research because it links research and intervention (concrete actions) and evaluates the outcome of an intervention so that lessons learned can be applied to future interventions. It is not just participatory research. It is participatory *action* research, which combines the top-down and bottom-up approaches.

Discussion and analysis are based on five years' experience (1993–1998) with a program in the Brazilian Amazon where PAR was adopted. The program is aimed at the stabilization of small farmers and the reduction of deforestation in a relatively large region known as Transamazônica.

Participatory Action Research

Action Research: A Brief History

Action research is generally considered to have been pioneered by Kurt Lewin in the 1940s. Lewin was involved in government programs to improve intergroup relations, and he sought methods of improving the contribution of science to the resolution of social problems (Lewin 1946). He realized that the only way for social science to advance was through research action in which experiments were carried out with natural social groups in their real-life environment. Lewin recognized that specific problems can be studied with small laboratory groups created for a given social experiment; however, he considered that the scope of such experiments is limited. This is because real-life behavior is determined by a variety of factors, ranging from the psychological and physiological factors of the individual level to cultural factors, such as the values and norms of a given society and the lifestyle of a social group.

When subjects of social experiments play an active part in the research by giving feedback to researchers through their actions and opinions, the action research is usually called participatory action research (PAR). PAR methods were further developed by sociologists, particularly in industry and management studies. In Norway, an ambitious program of action research on "industrial democracy" was launched in the 1960s by representatives of industrialists and trade unionists, who wanted to increase the worker's participation in management decisions (Emery and Thorsrud 1976). Research actions were conducted in many different industries over a period of 14 years. As a result, new forms of nonhierarchical work organization were developed and the Norwegian Parliament adopted new laws concerning the relationship between workers and capitalists. This effort contributed significantly to the development of the social-democratic experience of the Scandinavian countries.

During World War II, other interdisciplinary groups in the fields of psychiatry or social psychology started to use research action methods to solve problems such as the treatment of military patients and the survival of a psychiatric hospital deprived of food supplies. The works of these groups gave birth to the Tavistock Institute for Human Relations in England and to the Institutional Analysis, the type of action research currently used in France (Liu 1997).

Action research has made significant advances in social innovations. One example is the promotion of new forms of work organization with greater individual and team autonomy (Trist et al. 1963). Action research has also produced important new knowledge about the relationships between technology and work organization in industry (Woodward 1958) and of open processes in social change (Thorsrud 1972).

In the field of education, researchers from Latin America have developed a concept of socially engaged PAR (Fals-Borda and Rahman 1991). The origin of this school stems from the Paulo Freire "conscientization" method of the 1960s (Freire 1970). Its basic tenet is that subordination of the rural or urban poor in developing countries derives not only from their lack of access to capital, but also from their lack of access to education and information. However, they have their own popular knowledge, which should not be disparaged but recognized and reinforced through dialogue with modern scientific knowledge.

PAR is a way of organizing this dialogue with science and helping people become conscious of their limitations as well as their potential strengths. One of PAR's main drawbacks is that it takes considerable time, especially if the social groups concerned are among the poorest and most dominated, and therefore the least likely to perceive themselves as having shared interests (Wright and Nelson 1995). PAR is therefore seldom compatible with the researchers' or funding agency's limited time span. A different situation arises when the disadvantaged are well organized and have already initiated a program of self-help. In this case, they have a strong tendency to try to control the researcher's activities and sometimes refuse evaluations or abstractions that appear useless in relation to their strategies (Barbier 1996).

Another important problem with action research is that it can pose questions about publication and social property of scientific knowledge. Since conflicting groups are often involved in the generation of this knowledge, it can be used by one group against the other, or it can alter the scientist's relation with his or her "clients" and with the long-term objectives of action research. Ethical questions arising from such situations have to be answered on a day-to-day basis (O'Brien and Flora 1992).

Verspieren (1990), drawing on a vast experience of action researches conducted in the field of education in France, Canada, and Belgium, proposed a classification of action research programs based on the type of relationship established between the researchers and the social actors (participants).

In experimental action research, the initiative comes from the researcher, who considers that action research will permit him or her to conduct social experiments needed to test hypotheses. In this case, the local actors are invited to participate in the action research, but they have limited influence on the process.

In the institutional or engaged type of action research, the demand comes from a particular group of actors who have identified their problem and strategy. They feel that they need a researcher to help them analyze their problem and conduct their research with more insight and from a broader perspective. This creates a problem for researchers because their role is not precisely defined and the type of research they are supposed to carry out is not explicitly negotiated. Hypothesis testing and verification are usually not possible. Although this type of action research can be efficient for problem-solving and institutional change, it may not produce scientific results because activism may take precedence over critical observations.

In the systems action research, the objectives in both research and social change are negotiated between the researchers and the practitioners. During the first phases, they also formulate a common definition of the problem to be treated and of the hypotheses to be tested. These hypotheses generally concern the results of interventions (activities carried out to solve the problem). After a series of controlled interventions and observations, an evaluation of the results is carried out. A comparison is made between the researchers' analysis and that of the practitioners. After this, the hypotheses and sometimes the definition of the problem may be revised and the cycle repeated. It is an iterative process that continues until the problem is solved (or until the partners desist).

For Verspieren (1990), the systems type of action research maintains the better balance between research and action and between the interests of the local actors and the researchers. It can solve concrete problems and produce scientific results. Because the systems action research can be rigorously documented and analyzed, it also offers the best possibility of replicability. Of course, to obtain replicability researchers must describe in detail the general environment in which the action research has been conducted and the problems and limitations that they encountered during the process. Systems action research also has the advantage of training practitioners in research methods, including competence in social practice, cooperative action, and analysis (Lewin 1946).

Applications of Participatory Action Research for Natural Resource Management

Ecosystem Management

Because of its complexity, ecosystem management has some of the characteristics of action research (Jordan III 1987). Ecosystem management cannot use the classic experimental approach because exact repetitions and controls are impossible. The efforts to model whole ecosystems based on a detailed description of their parts and interaction has proved to be inefficient (Jordan and Miller 1996). However, it is possible to use "normic statements" (statements about the average condition; Pomeroy et al. 1988) to predict, with a reasonable level of certainty, certain characteristics of ecosystems observed from a holistic perspective and therefore to design management alternatives and test them by the old trial-and-error method. There will always be a level of uncertainty about the final result, but this adaptive management is better than no management at all or management through random interventions.

Environmental Problem-Solving

Local environmental questions can often be described as natural resource management problems. At the level of individual property, the improvement of natural resource management is a question usually treated by agricultural sciences, forestry science, and so on. The natural resource management analysis takes as basic elements the natural resources at stake and the stakeholders who have some interest—directly or indirectly—in the use or maintenance of these natural resources. It is not an environmental problem as such.

It is characteristic of an environmental problem that the practices of individual actors end up affecting other actors, either engaged in the same activities (in a problem of a common-pool resource) or engaged in other activities (in this case, a situation in which various stakeholders have different interests in a common resource).

The idea of using action research in the field of environmental problems is relatively new (Castellanet 1992, Jiggins and Roling 1997). Its first premise is that purely technical solutions are seldom adequate to solve environmental problems. Most require complex negotiation and agreement. Pragmatic experience has shown that the (subjective) understanding of the different stakeholders is attained more rapidly when research on natural resource management is integrated into an action research approach.

Farming Systems Research and Development

Although the objective of the project described in this book has been to improve the use of the forest resources in the Transamazônica region of Brazil, the group selected for primary cooperation, for reasons discussed in Chapter 5, was a farmer's organization. Although agricultural research and development is not usually considered part of natural resource management, it is strongly related both in content (each farmer manages natural resources on his farm as part of his activity) and history.

The importance of farmers' participation in research aimed at improving agricultural production has been discussed since the 1970s in International Agriculture Research Centers (Rhoades and Booth 1982). Recently, participatory research has become a fashionable concept among agronomists and socioeconomists working on the design of development projects (Chambers et al. 1989, Scoones and Thompson 1994). It is also recommended for the field of forestry by the Food and Agriculture Organization (FAO) and the Overseas Development Institute (ODI; London) in the form of rapid rural appraisal methods to foster participatory community forestry.

In the field of agriculture, farming systems research and development methods have involved very different levels of farmers' participation, ranging from simply integrating farmers into the study as research objects to their effectively controlling the research process (Pillot 1992, Ashby 1986). When farmers effectively become real partners in the process and effectively change the research agenda, one can speak of "research-action for agricultural rural development."

Commonalities Among Approaches

The different schools of action research can be characterized by different levels of involvement and commitment of the researchers with the target groups, different levels of participation, different research methods, and different disciplinary fields. However, they all share a critical view of the capacity of mainstream science to solve social problems and a belief that action research is more efficient for provoking social change. The action research schools also believe that action research can contribute significantly to the progress of social sciences and interdisciplinary problems such as those of the environment and development.

Lewin (1946) suggested that researchers can learn much more by intervening into reality than from only observing it from the outside. To test this idea, he designed sociopsychological experiments with social workers who then went back to their communities, observed the results, and evaluated them. Through this experience, he felt that action, research, and training would grow together.

More recently, Liu (1990) again argued that intervention through action research in social sciences is not only a powerful heuristic tool, but also presents an opportunity to empirically confirm the findings of research through real-life experimentation. This avoids the situation in descriptive social science, in which predictions can be made but experimentation is generally impossible. And fortuitous observations are rarely sufficient to confirm these predictions, since social events do not obey researchers to confirm their predictions. Anthropologists and sociologists also have observed that experimental modifications or disruptions of social systems that may result from planned or unplanned interventions are often more revealing than long, neutral observations.

Authors of radically different backgrounds such as Freire (1970), Liu (1992), and Jiggins and Roling (1997) have noted that all types of action research promote individual and collective learning and increase individual autonomy and problem-solving capacity in a durable way. Although all types of action research are "participatory" in nature, we have chosen to use participatory action research to describe the method used in this book, first, because it relates to the Latin-American historical and political context and, second, because it insists on the importance of local actors in the process.

Knowledge Needed for PAR

When scientists try to solve a complex problem, they need to use two types of knowledge (Lewin 1946). The first type is derived from traditional academic science. It concerns laws and regularities that can serve as a guide for achieving certain objectives under certain conditions. But researchers also need to know the specific character of the situation at hand. Therefore, they must conduct a diagnosis of this situation (the other type of knowledge). One can also say they need a model of the complex object on which they want to intervene. But no diagnosis can be perfect the first time. Action research is a way of testing this model and improving it gradually. It also is the best way to test any proposed method of intervention because it evaluates results against original hypotheses. By this means, errors can be observed and corrected, and new questions and hypotheses can be formulated (Verspieren 1990).

Systems Approaches

The study of complex systems requires special methodological tools (heuristics). The debate among holism, mechanical systems, and reductionist approaches in biology illustrates this problem. The history of natural sciences demonstrates that biologists had to modify their research strategies progressively since the end of the 19th century, from the search for simple linear mechanisms as causes of all observed phenomena to a more sophisticated approach that considers living organisms as complex systems. (See the history of research on alcoholic fermentation in Bechtel and Richardson [1993] or the debate on population genetics described by Wimsatt [1980].)

The formalization of the systems approach was greatly facilitated by the development of digital computers in the 1950s, which permitted the development and use of systems models by nonspecialists (Simon 1981). Today, ecology makes wide use of the systems approach, not only in ecosystem ecology, but also in population genetics and evolution (Levins and Lewontin 1985) and in human ecology, which has widely adopted the hierarchical approach (Young 1992). All recent research on the biosphere and global change has a strong systems component, which is necessary for modeling. In agronomy, the systems approaches have been incorporated into farming systems research (Shaner et al. 1982) and the hierarchical approaches for agro-ecosystems analysis (Conway 1985).

Unfortunately, the scientific tradition favors the mechanistic-experimental approach and considers systems theory to be a particular branch of mathematics rather than an essential tool of naturalistic investigations. As Bechtel and Richardson (1993) point out, we have severe limitations in our intuitive understanding of systems, which also explains our natural inclination toward simple localization, mechanical explanations, and reductionist approaches. As a consequence, systems approaches and modeling of complex objects are viewed with suspicion in the dominantly reductionist scientific community. Many scholars, despite the pioneering work of Malinowski (1949), Parsons (1951), and more recently Morin (1991), also see use of the systems approach in social sciences with suspicion. On the one hand, they criticize its lack of precision and the tendency to use the "systems" language as a new jargon with little precision (Olivier De Sardan 1996b); on the other hand, most of these scholars are afraid of the mechanical and quantitative aspects of systems models. Many of these criticisms are unfounded, but the fact is that the systems approach is as yet seldom used, at least explicitly, by social scientists.

A field in which the systems approach is commonly used is the "sociology of organizations" (Simon 1973, Checkland and Scholes 1990, Crozier and Friedberg 1977). However, the soft systems approach defined by these authors is quite different from that used in the biological sciences. The soft systems model was developed by a group of researchers involved in industrial management analysis. Checkland and Scholes observed that the classic (mechanical) systematic approach (quantitative modeling) was giving good results when applied to the optimization of predictable activities such as building a factory. However, it could not be applied to human organizations, since meaning and values are essential in understanding and predicting human behavior. So, instead of assuming that a given organization can appropriately be represented as a machine, they proposed to compare the actual functioning of these organizations from the point of view and objectives of their participants (actors). Hence, the term "soft system." The researchers are using nonquantitative models of human activities as diagnosis and communication tools aimed at improving the efficiency of the organization studied. The basic principle of the soft systems methodology is that practitioners should avoid considering that the real social world can be described completely by a system: in examining real-world situations characterized by purposeful action, there will never be only one relevant holon (systems model), given the human ability to interpret the world in several ways. It is therefore necessary to create several models of human activity and to compare them with the real world, especially through the observation of the results of planned interventions (action). The soft systems methodology is, therefore, a form of systems action research on social objects and problems. A similar approach has been used with some success in the Man and the Biosphere program of Obergurgl (Moser and Peterson 1981) and in the Vosges research program in France (Legay and Deffontaines 1992).

Hierarchical Systems

Today, the analysis of complex systems models is based on the hierarchy theory (Allen and Starr 1982). We can analyze the behavior of systems that are composed of subsystems, which are themselves composed of subsubsystems, and so on. Depending on the amount of interactions between hierarchical levels, compared with interactions with other subsystems, the system can be considered strictly hierarchical (hierarchically decomposable) or only approximately hierarchical. Although hierarchization is a general phenomenon in living organisms, in many cases the systems observed are not strictly hierarchical (especially at the level of populations, ecosystems, and similar systems) and the determination of an a priori hierarchy is sometimes arbitrary.

In ecology, hierarchical concepts have been adopted by ecosystem ecologists (Allen and Starr 1982), human ecologists (Young 1992), and landscape ecologists. Farming systems research is a good example of the use of an approach that uses hierarchy theory. Initially, the level of investigation was the farming system, composed of the farmer and his or her farm (factors of production) within a relatively short time scale (a few years needed for technology adoptions). However, it quickly became clear that this scale was too limited to completely explain farmers' choices. Farming systems can be decomposed into lower level subsystems (the cropping system and the livestock system in particular), which themselves can be decomposed into sub-subsystems (e.g., the plot, the soil-plant interaction). At a higher level, farmers' strategies are frequently based on long-term perspectives for their children and the farmers' old age. Farmers interact significantly with other farmers (the agrarian society), exchange labor or other factors of production, and are constrained by the rules of society, particularly in the market situation. They also are influenced by land tenure rules, religious or other social obligations, and available local knowledge (Conway 1985).

At a still higher spatial and temporal level, we find the traditional discipline of "comparative agriculture," which tries to reconstruct the history of the relationship between agricultural technologies and societies. In Europe, comparative agriculture was studied initially by historians such as Bloch (1966), who analyzed the agricultural revolutions. Boserup (1965) studied the relationship among population growth, technological change, and agricultural intensification. Ruthenberg (1980) analyzed the history of farming systems in tropical areas. This type of analysis is based on the assumption that there is a strong link between agricultural productivity, the economic and political organization of the society, and its dominating ideology. The hierarchy theory certainly constitutes a useful framework for practitioners of farming systems research and development. This helps to identify some of the strong structural constraints in the agrarian society that are not likely to change in the near future, but have to be taken into consideration in all development programs and in national policy design (Mazoyer 1986).

The study of natural resource management proposed by Ostrom et al. (1994) uses a hierarchical systems approach. The individual stakeholder's decisions depend on the group's rules and norms, which in turn are influenced by national laws and institutional environment. Reciprocally, the groups' rules are designed and enforced by the individuals themselves, and state laws are enforced by the groups.

An important implication of systems theory is that it demonstrates that the behavior of a nearly hierarchical system is only approximately predictable. Weak interactions of subsystem components with outside entities or other subcomponents create variability in the responses, which are equivalent for all practical purposes to stochastic variations. A general point of agreement among systems theorists is that the existence of hierarchies in nature clearly brings the question of the relevant level of investigation into research. Contrary to the radical reductionist claims, it is not only useless but also futile to try to explain what happens at the population level by means of fundamental laws of physics (Levins and Lewontin 1980). But this does not mean we should abandon reductionism and rely on metaphysical holism. Many scientists now advocate a complex reductionist research strategy. Complex reductionism is based on the following principles (Wimsatt 1980):

- Scientists may legitimately study regularities in the behavior of a class of systems without necessarily looking for their explanation at the lower level.
- Regularities at a certain level of systems hierarchy become mechanisms at the higher level.

In terms of heuristic strategies, this implies that research should work on attaining a generalist diagnosis of the system and on testing of proposed mechanistic relationships simultaneously. This means a constant to-and-fro movement between systems modelers, who formulate hypotheses (based on observations) on how the system may work, and experimental studies and observations, which confirm or falsify these hypotheses.

Is Participatory Action Research Scientific?

PAR breaks with the traditional view of a research program, at least from the point of view of research agencies and academic tradition, in which researchers first define their subject, then narrow it progressively until they find relevant mechanisms to analyze. The participatory research agenda cannot be planned in advance because it must be renegotiated with the end users periodically. Consequently, many scientists still consider that action research, or participatory research, is not a legitimate scientific approach. Various criticisms have been made from epistemological and sociological points of view. These criticisms concern the lack of objectivity of PAR, its applied character, its complex and interdisciplinary objectives, and its frequent lack of institutional recognition.

Participatory Research and Objectivity

A common criticism of participatory methods and action research is that the researcher cannot maintain an objective stance, since he or she becomes involved emotionally and socially in the problem being studied. This problem of participants influencing observations has long been recognized in the social sciences (Malinowski 1949) and has also been demonstrated in pure physics. In fact, the Heisenberg theorem states that neutral observations are impossible because the observer always interferes with the observed object.

Defenders of PAR consider that researchers must recognize their subjectivity and use their inner capacity to distance themselves from the action and the object being studied. Risks of subjectivity are increased, of course, when personal involvement increases. But practitioners usually consider that (1) the gains in better understanding and insights into other actors' logic can largely offset the potential disadvantage of increased subjectivity and (2) a conscious researcher can learn to avoid the biggest pitfalls of subjectivity and intellectually separate the moment at which he or she participates in the action as any local actor from the moment he or she analyzes and thinks about his or her observations. This normally takes place when the researcher is writing field notes, listening to recordings, and writing papers and discussing them with critical colleagues.

Falsifiability

A hypothesis can never be proven true under all circumstances. Some condition always remains that is not yet tested, under which the hypothesis can be false. However, for a hypothesis to be considered legitimate, it must be falsifiable; that is, it must be possible to devise a test that can theoretically prove the hypothesis to be incorrect. Experiments in the social sciences have not been able to pass this test of "unfalsifiability" (Popper 1983).

It is doubtful that general laws, applicable in all places and at all times, will ever be discovered in the social sciences. One reason is that humans are able to change their behavior when they see an advantage in acting contrary to general laws that describe the behavior of the majority. Such instances are particularly common in economics. Therefore, most social researchers consider that the epistemological criteria of the social sciences cannot be the same as those of the physical sciences. Explanations of social behaviors should be based on an effort to comprehend interpersonal and intercultural communication, rather than statistical inferences. In other words, the social researcher's work is based on logical inference, not statistical inference (Mitchell 1983).

Replicability

Most social experiments are impossible to replicate. However, observations based on models and processes must have some level of replicability and generality to be useful to other practitioners. As Verspieren (1990) notes, this supposes that when the results of research action are presented and published, a special effort must be made to describe the general conditions (environment) in which it was conducted, and to analyze in detail the process through which it occurred and the crucial decisions that were taken. After this effort of objectivity and generality is made, the results of the research action may be transferable to similar conditions and problems. The type of predictability is obviously not absolute. We have to face the "relative uncertainty" of predictions, which still is much better than no prediction at all.

Applied Research versus Pure Science

Although there are studies of social behavior that can be considered basic or fundamental, the social problems dealt with in this book are *applied*; that is, the research is directed toward solving a particular social problem. Many scientists today consider applied science to be an inferior form of science. Some philosophers justify this opinion by stressing that the fundamental aim of science should be to acquire knowledge and understanding of the world, not social or moral objectives (Musson, cited by Schaffner 1992). This, like all definitions, is arbitrary and is also an ethical judgment.

Science can also be considered as an institution of modern society, whose objective is to systematize acquisition and transmission of knowledge in order to improve its control of the material world. This instrumentalist perspective was clear for the first empiricists such as Francis Bacon. Therefore, to find out whether the discrimination between pure and applied science is justified, we need to look at the type of knowledge and methods of applied science, rather than its ultimate aim.

There is nothing a priori in the methodology used by applied science that differentiates it from pure science. Problems are defined, hypotheses formulated, explanations sought, and confirmation obtained inductively through observations or experimentation—exactly as in pure science. The main difference lies in the choice of the problems investigated. Pure science tends to work on the problem puzzle, defined by its research program (Kuhn 1977), whereas applied science selects problems based on their social use. However, the difference between the two is often more psychological than real, given that most research funding is decided by governments or private firms based on expected economic or political products. Positivists argue that only pure research can build theories and find general laws of nature, whereas applied research has no such objective. Therefore, only pure science permits fundamental progress in knowledge (which, in the positivist view, is the driving force behind human progress). Salmon (1992) claimed that scientific knowledge is the supreme achievement of our society. This claim is highly debatable. Some historians, but also physicists (e.g., Duhem 1914), have demonstrated that the progress of pure science largely depends on the progress of technology for indirect observations and measurements. The fact that technology progressed before science has been demonstrated by Kuhn (1977) for the 18th and beginning of the 19th century. Pasteur's discoveries in microbiology were also a consequence of his work for the wine industry (Bechtel and Richardson 1993). Quite appropriately, Pasteur declared that there is no such thing as pure science and applied science. There is only science, and the applications of science. Even today, basic science and applied science are hardly distinguishable. Digital computers were initially developed in the 1940s within a typical applied research program with military objectives, but their impact on scientific development has been tremendous.

In spite of the difficulty of separating pure from applied science, the concept of the superiority of pure science remains deeply rooted in the scientific community. It can probably be explained largely as being the result of the triumph of "scientism" in the never-ending war for public funding and recognition. Scientism can be thought of as an ideology that justifies the material sustenance of its advocates (the scientists), just as religions feed their priests based on theological arguments (Feyerabend 1975). Auguste Comte was not so foolish after all when he proposed the institution of a scientific religion (religion of humanity) in the positivist State (Bryant 1985)!

We need other criteria to characterize an approach as scientific or nonscientific. We have to be able to distinguish scientific approaches from religious, magical, or even pragmatic approaches. For example, a Kayapo Indian might know a lot about the forest and use this knowledge aptly to survive, but it would not be considered science. If it were, there would be no limits between science and common knowledge.

Science as an Institution

Another way of characterizing science is to use a socioanthropological approach. In this view, scientists are part of research institutions, with defined rules and functions (Latour and Woolgar 1986). Science is simply the activity of these recognized scientists, whether or not it follows specific epistemological criteria. The problem is that such a definition eliminates the possibility of independent researchers, or "marginal" scientists, probably few in number but nevertheless present.

Action research and interdisciplinary studies are not within the present dominant normal paradigm and therefore are not easily accepted within academic institutions, which are marked by conservatism and respect for the accepted norms of scientific production and status. But, as Kuhn (1977) has abundantly demonstrated, the marginal scientists of today may become the normal scientists of tomorrow, which is a fundamental form of scientific progress.

The Scientific Output of PAR

An important criterion for judging whether an activity is scientific is the product of that activity. Science is the social activity that results in a scientific product, and this product that consists of written material submitted to peer review. Scientific activity in the latter sense can therefore be characterized as the production of written documents with specific characteristics. These documents are circulated and criticized within a certain community to characterize living science. This peer review system guarantees that basic criteria such as objectivity, clarity, and coherence of analysis are actively pursued by researchers of the same "school."

PAR papers are difficult to publish in journals specializing in reductionist science. In such journals, the focus is on mechanisms rather than performance of the system. Acceptance is rare, partly because such papers depart from the traditional positivist criteria for science and partly because the active scientific community interested in these methods is still too small to launch its own journals. However, as in the research reported in Chapters 6 and 7, documents are produced (see Appendix 2) that have the basic qualities of scientific documents (such as research objectivity, separating observations from interpretation, and analysis).

Conclusion

Although we admit that PAR is not included in the dominant science paradigm, it cannot be discarded as a scientific approach on the basis of objectives or on institutional considerations. PAR takes its legitimacy from the fact that it treats problems that normal science does not know how to handle.

3

CHAPTER

Conceptual Framework

The Interpretationist Tradition

There are two fundamentally different and conflicting views of social science: the *naturalist tradition*, in which the objective is to understand the cause behind human activities, and the *interpretationist tradition*, which holds that human action is the result of intentional choice stemming from the conjunction of desires and beliefs with external circumstances (Rosenberg 1988).

Some natural resource management practitioners have adopted the marginalist economic theory, which is a mild form of naturalism. Marginalists try to explain individual economic choices by using a rational calculation based on the maximization of quantifiable "utilities" derived from goods acquired or services obtained by the individual agent. However, they have to recognize that utility is hardly measurable, since preferences may change in relation to circumstances and time. In fact, utility reflects only the subjectivity and variability of desires and beliefs.

Because econometric approaches to farmers' decisions based only on observable external indicators have not been very successful, many practitioners have adopted the interpretationist paradigm, using the "rationality hypothesis" as a central tenet in their analysis of individual action. This hypothesis can be formulated as: If X (individual) wants D (desired object) and believes that A (action) is the most appropriate way to obtain D, then (all other things being equal) X will do A.

From the positivists' (behaviorists') point of view, desires cannot be experimentally measured unless one precisely knows what the beliefs are, and vice versa. So the interpretation exercise always remains subjective and its premises are not testable. For the disciples of Popper (1983), this "unfalsifiability" disqualifies interpretationism as a scientific method. Indeed, this confirms the common view among natural scientists about social science: it is simply not a science. There is no Nobel Prize in social science.

However, a certain amount of predictability can be obtained from an interpretationist approach. Interpretationism often obtains the same level of accuracy as folk psychology when it tells us, for example, that "the probability of an average American car driver to ignore a red light is low." In fact, some descriptive anthropologists believe that their goal is attained when they can achieve the same level for predicting social behavior as members of the cultural group they are studying. Their job is really to learn the *rules* of a given group, starting with the rules of meaning (the language) and including beliefs, desires, and social rules (Winch 1958).

In farming systems research programs, farmers' decisions could usually be analyzed in the interpretationist tradition, as a result of rational choices, although these choices were not always clearly stated by the farmers themselves (Rhoades 1986). This, of course, is not in itself proof of the validity of the theory; there is always a danger of circular reasoning and ad hoc explanations. However, in many cases, the reasons for adopting or not adopting a new agricultural technology were observed to be technoeconomic in nature and, therefore, could be somehow quantified in contrast to strictly cultural reasons. Many examples of this type can be found in the literature (Collinson 1983; Rhoades 1984). The fact that farmers' technical choices often refer to economic considerations rather than magical or religious ones can possibly be explained (in a teleological sense) by the fact that successful agriculture is a basic material necessity for the survival of farmers and their cultures. Therefore, farmers' know-how and methods are essentially aimed at practical and economic results. Based on these observations, researchers in agronomy who adopted the systems approach have proposed various models of the farmers' decision-making process, in which long-term objectives and strategies interact with short-term decisions (tactical choices). Decisions result from the confrontation between a farmer's observations and his expectations, which are based on personal experience and acquired knowledge (Brossier et al. 1990). A broad model of individual knowledge and decision systems is represented in Figure 3.1.

We take, as an additional hypothesis, that other forms of subsistence, such as extractivism, fishing, and hunting, or extensive cattle-rearing, also can be best understood according to the technical and economic

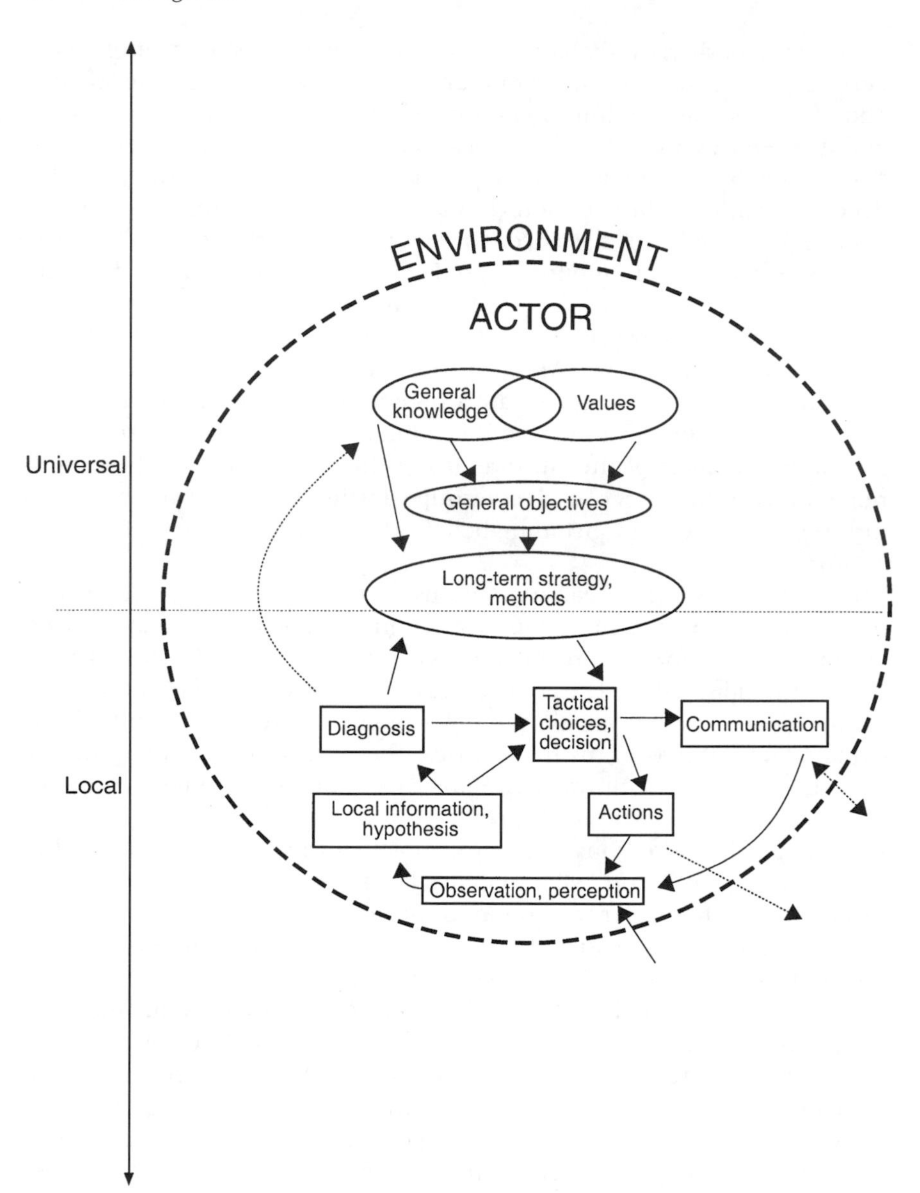

FIGURE 3.1. Model of individual knowledge and decision system.

background of any given cultural group. Economy here is taken in its broadest sense, as used by economic anthropologists such as Sahlins (1989) and Meillassoux (1975). Economic value is never absolute. It is completely linked with cultural values, local custom, and laws. Therefore, it is relative to a given society and, within this society, to a given status group with a same "habitus," as coined by Bourdieu (1994).

Social scientists who study social change maintain that one has to consider the total set of interactions between the different actors in a given arena around specific targets to understand how innovations are introduced in the social system, how new resources are divided and used, and so on (Olivier De Sardan 1995). Giddens (1979) and Balandier (1967) have defined this as "socioanthropology of development." The basic tenet of development socioanthropology could be: "Within the limits of information, uncertainty and other constraints (e.g. physical, normative, socioeconomic), social actors are competent and capable" (Long and Long 1992, p. 23). This means that they act rationally in the sense defined by the interpretationist paradigm. What is needed is to develop theoretically grounded methods of social research that allow for the elucidation of actors' interpretation and strategies, and how they interlock through processes of negotiation and accommodation. Unfortunately, researchers still commonly give priority to the forms of logic that superficially appear most relevant to the type of study undertaken—economic logic in the study of production strategies and symbolic logic in the religious field. But human behavior cannot be adequately described under one form of dominant rationality, whether economic, cultural/symbolic, or political (Giddens 1979, Olivier De Sardan 1995). Economic logic *also* plays a role in rituals and symbolic logic *also* determines economic behavior.

An Interpretative Model

Figure 3.2 is a model of the interaction between farmers and researchers. Each of the two partners can be characterized at any given time by their long-term objectives, values, knowledge, and strategy. Knowledge might be divided between value-laden and objective knowledge, but it is not always possible to separate the two, even when analyzing scientific choices (Vietor and Cralle 1992). We can also distinguish between general knowledge (applicable everywhere) and local knowledge (applicable only in the local or regional context). Our research program was aimed at understanding and solving environmental problems at the level of the Transamazonian region, thus most of the information gathered and analyzed was local in nature. This local knowledge was used as a basis for a regional diagnosis, which served as the basis for a long-term strategy. The diagnosis changed and improved as new information was incorporated into the knowledge system. In fact, this evolution of the diagnosis—hopefully toward a greater accuracy and realism—was one of the main aims of the research team.

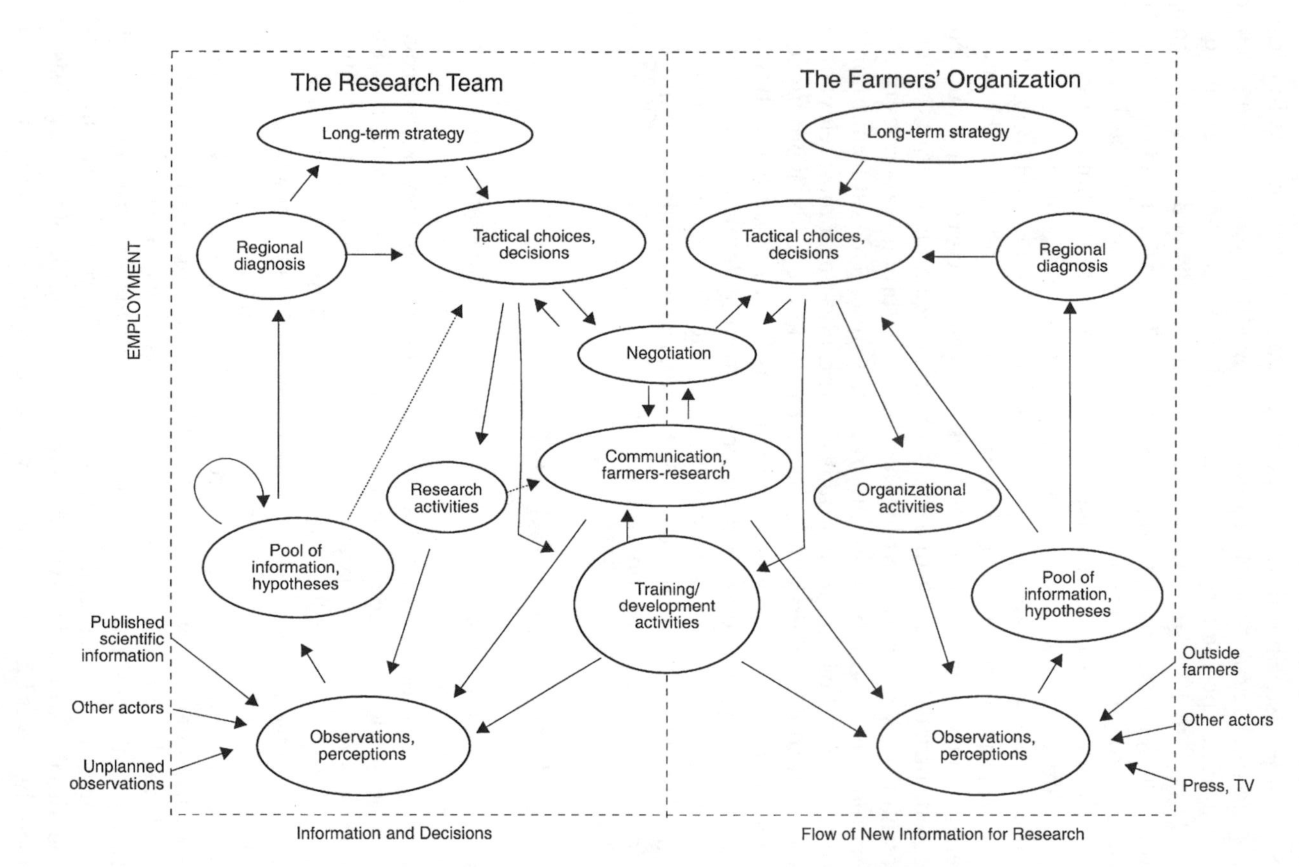

FIGURE 3.2. Model of interaction between the research team and the farmers' organization.

In contrast to the research team, with its well-formulated and structured diagnosis, the farmers' organization did not put a priority on formalized diagnosis. Theirs was a more informal and less openly discussed world view.

The Extended Rationality Postulate

Since we are studying not only exchanges of information between farmers and researchers (level 1), but also how these exchanges affect their behavior (level 2), we need to adopt the rationality hypothesis previously described. This, however, may not reveal long-term goals, since such goals may not be apparent from conversations or short-term actions. We therefore adopt an auxiliary postulate: X can formulate a diagnosis (based on his or her knowledge of the situation and hypotheses) and a long-term strategy so that: if X wants LO (long-term objectives) and believes, based on his or her diagnostic (D) that the best way to reach LO is to obtain O (immediate objective), then X wants O.

There is a great difference between the extended rationality model and other models of human communication, such as artificial intelligence (Simon 1981), technology transfer (Roger and Kincaird 1981), knowledge systems (Roling and Engel 1992), and other theories of indigenous knowledge commonly used in ethnobotany and other ethnosciences. Here, we consider that human communication is not a pure transfer of information between two persons, but always involves cultural values and emotional aspects. That culture and emotion are an essential part of communication is demonstrated by the failure of the neopositivist school to translate human language into an objective machine language. The meaning of words is always relative to a given cultural environment and also to a given shared experience between two interlocutors. Even in physical science, no experiment can be conducted without using relative concepts, which are themselves "unverifiable" (Duhem 1914).

The Constructivist Model

A constructivist model incorporating a systems representation of reality is the approach taken here. It is a paradigm within the broad interpretationist school. Constructivism suggests that internal coherence, the ability to communicate and cooperate with the different social actors, and pragmatically contributing to a solution of problems should be some of the criteria of validity. The constructivist model is a way of formalizing the interaction between human beings in general. It was first used by Piaget (1972), who studied parent-child relationships and the resulting building-up of capacities and personalities of the children. In our case, the relationships are between researchers and other social actors with whom they are strongly linked and on whom they conduct research. Constructivism permits a description of how researchers progressively construct their own frame of reference, based in part on scientific tradition; how, at the same time, farmers are affected by researchers' activity, directly or indirectly; and how this interaction translates into changes of attitude and practice.

Le Moigne (1984) describes the basic principles of constructivist epistemology as follows:

- A thinking subject is able to construct representations of the world, based on his or her particular experiences of his or her interactions with perceived and conceived phenomena.
- Different representations of the world can be constructed by the same subject (or different ones), all of which are legitimate in their particular context.
- Constructivist epistemology therefore abandons the idea that there is only one real objective description of the world, as is characteristic of positivist epistemologies. Constructivist epistemology also departs from the positivist tradition of the "objective and outsider" observer of the reality.

Various students of action research and of the systems approach have observed a natural tendency of action research practitioners to use constructivist paradigms and a systems approach (Avenier 1992). Action research method is based on the organized interaction between researchers and social actors and presupposes a simultaneous change in both sides. The constructivist epistemology, which takes as a starting point that all knowledge is constructed within a given society, is better adapted to this situation (Le Moigne 1984).

Case Study Methods

The case study is a multimethodological approach, which keeps a certain equilibrium between theory and empiricism and between deductive and inductive methods (Stoecker 1991). Frequently, a case study based on an exception to the normal situation has more explanatory power than an observation of normal situations. Very often in anthropological studies, crisis situations reveal more about the social system than normal ones (Mitchell 1983).

Case studies are important for observing the limits of accepted theories or models and for proposing new elements to be incorporated into them to improve their generalizability. Research action can be considered a special "limit" type of case study due to its duration and the implication of the researchers in the action, and of the participation of local actors in the research. Research action has several important advantages over conventional case studies: the local actors contribute continuously to the verification of research hypothesis and models, and it permits the researchers to realize social experiments that are both ethically acceptable and epistemologically efficient (Liu 1997).

Case studies have made important contributions to social sciences, especially in anthropology, but are often criticized as being nonscientific. The question of their "representativity" regarding the broader universe is often questioned. But Mitchell (1983) believes that the validity of the case study is not based on quantitative or statistical inference, but on the coherence of the causal or logical inferences that can be derived from it. The heuristic value of a case study therefore depends on the fact that the set of events analyzed within the particular context of the case study can explain a new relationship between facts and social traits within a given theoretical framework.

What is observed and analyzed and how it is observed within the case study largely depends on the assumptions and theoretical framework of the researcher. However, if the phenomena observed do not fit the theory precisely, the researcher must be able to test other possible explanations to correct or substitute the initial hypotheses. The case study is a multi-methodological approach, which keeps a certain equilibrium between theory and empiricism and between deductive and inductive methods (Stoecker 1991).

We believe, like Olivier De Sardan (1995), that in the relatively near future systematic comparisons between a significant set of case studies focusing on the same theme will become possible, yielding new discoveries and conclusions. Such themes may include the interaction between outside researchers or "developers" and local populations in the context of external intervention and the resulting social changes and adoption of innovations. These conclusions, in turn, will help us to better prepare and follow up these interventions and to better train the new professionals who will be needed to approach these new challenges (Roling 1996).

4

CHAPTER

The Resource Management Problem

Tropical Deforestation

Tropical deforestation has been accelerating since the 1960s. Between 1980 and 1990, an average 15.4 million hectares per year of forest have been cleared, or 0.8% of the total forest area (World Resource Institute 1994). One third of this total (4.6 million hectares) is rainforest, which lost 0.6% of its total area per year. However, the total forest area affected by human activities is much larger than the deforested areas. During the last 30 years, logging and wood extraction have grown steadily, as determined by area logged and volume extracted. New forest plantations are insufficient to compensate for deforestation. They represent only about 1.8 million hectares per year and are mostly monospecific and exotic fast-growing species (FAO 1995).

Although some previous, alarming estimates were exaggerations, many tropical forests will, no doubt, disappear or be severely degraded in the next 20 years, especially in Central America, West Africa, and Southeast Asia. With recent technology, we have demonstrated that we are able to destroy the natural forest at a pace never before observed in history. International concerns, originating in great part from scientists and then popularized by the environmentalist movement, have predicted that this deforestation may have profound consequences for mankind.

Concerns have been expressed about the impact that massive tropical deforestation would have on the local and global climate, through changes in reflectance, evapotranspiration of the vegetation as well as increased erosion and changes in the water cycle (Salati 1990). Increases in atmospheric carbon also exacerbate the greenhouse effect. For each hectare deforested, an estimated 130 tons of carbon go into the atmosphere. The conversion of tropical forests is responsible for an estimated 26 to 33% of present carbon dioxide production (IPCC 1990).

It is commonly thought that tropical forests, especially rainforests, contain more than half the world's biodiversity. Based on the species-area model of biogeography, Wilson (1992) calculated that a reduction of half of the remaining areas of rainforest should cause the extinction of 10 to 22% of the species within. This means that with a conservative estimate of the rainforest biodiversity at 10 million species, 27,000 species are doomed to extinction each year if deforestation continues at its present pace. This is an unprecedented impact, and numerous ethical and economic arguments have been made that such destruction is not justified in terms of the resulting economic growth.

The rate of recovery through natural succession depends on the level of perturbation initially applied. Cases of irreversible degradation seem more rare than is commonly believed (Buschbacher et al. 1988), but even if deforestation were to stop today, the return to the initial level of biomass and biodiversity would take centuries (Saldarriaga 1988).

Deforestation in the Brazilian Amazon

The Amazonian forest (Map 1) is by far the largest existing tropical forest. In Brazil alone, it covers 4,090,000 km^2 or 31% of the world's total forests. Huge areas are deforested annually at rates estimated to reach 80,000 km^2 per year. This represents half of the total world tropical deforestation (World Resource Institute 1990). An increase in the annual rates occurred between 1950 and 1990. Based on satellite photographs, Mahar (1990) showed that the area of deforestation in the Brazilian Amazon represented 0.6% of the total forest area in 1975, 2.5% in 1980, and 12% in 1988. This means that annual rates of deforestation grew from 0.4% per year in 1975–1980 to 1.2% in 1980–1988.

Alarmist forecasts resulted from the extremely rapid deforestation observed in the state of Rondonia after the opening of a new highway between Cuiabá and Porto Velho. More accurate estimates made by the Brazilian National Institute for Space Studies, based on Landsat satellite images in 1989, showed that the total area deforested was in fact only 8%, giving an annual rate of 0.6% in 1980 to 1989. Other estimates indicate

Map 1

that deforestation has been actually closer to 15,000 km^2 per year between 1978 and 1988, representing an annual deforestation rate of 0.4%.

Establishment of crops and pasture were the main reasons for deforestation. Area under cultivation and cattle stock both increased quickly between 1970 and 1980 (11% per year in the area cultivated, 9% per year in pasture), then slowed down in 1980 to 1985 (3% in cropping area, 6% in pasture). In 1988, an estimated 22,000 km^2 of Amazon rainforest had been converted to crops and 133,000 km^2 were in pasture. Most pastures go into a degradation process within 5 to 10 years (Hecht 1984, De Reynal et al. 1995). Estimates of the area severely degraded ranged, in the early 1980s, from 15% (Toledo and Serrão 1982) to 50% (Hecht 1984). Note, however, that a degraded pasture is not necessarily a barren ecosystem: on the contrary, most degraded pastures are invaded by woody shrubs.

Crops and pasture are not the sole reason for rainforest exploitation. Various levels of human intervention result in a diversity of perturbed ecosystems: exploited forest where valuable tree species have been extracted, but the structure and function of the forest are not severely affected (Johns 1988); agroforests created or encouraged by humans with various levels of internal biodiversity; pure stand forest plantations; diversified agriculture within the slash-and-burn system; seminatural pastures; and so on. The area of forest degraded or affected by human activities is estimated to be 2.5 times higher than the deforested area (Skole and Tucker 1993).

The economic growth of the Brazilian Amazon was made possible by heavy fiscal transfers from the federal government (World Bank 1992). Schneider (1995) observed that federal transfers in relation to the regional gross national product were reduced significantly between 1975 and 1985, from 24 to 10%. Its decrease was more pronounced in Pará (21 to 6%) and Rondonia than in other Amazonian states. Nevertheless, the general impression remains that large areas of forest have been destroyed for meager economic and social results and that, even from a strictly economic point of view, government investments would have been more efficient in other sectors.

The Target Groups

Authors who studied the causes of deforestation in Brazilian Amazonia generally agreed that the idea of reversing the present trend is not a lost cause. In other tropical countries, deforestation is the result of high demographic pressure; agricultural land is scarce and the major part of the population still depends on agriculture or wood extraction for a living. This is not the case in Brazil, however. Although the population in Amazonia has increased, most of the increase seems to be due to local demographic pressure, since the net migration to northern Brazil was estimated at only 770,000 between 1970 and 1980 (World Bank 1992).

The Amazonian population remains relatively small, of low-density, and rural, compared with most other tropical countries. As a result, Amazonia does not have powerful demographic, economic, and political forces combining to press for more deforestation (Sawyer 1990). The rapid Amazonian deforestation has been more the result of federal government incentives and investments.

Today, the federal government is inclined to reduce its expenditures in Amazonia rather than make new investments. Therefore, a reasonable possibility is that a more sustainable model of developing the region can be implemented without additional deforestation. For this, more appropriate national policies should combine with the interests of the majority of the local people, small farmers, extractivists, fishermen, and indigenous populations to encourage more sustainable forms of agriculture and extractivism. The main problem is to get local support for these alternative policies and to avoid the local elite's efforts to maintain the status quo (Schneider 1995). Many scientists are certainly willing to contribute to this new model of development, but they have difficulty in deciding how to do it most efficiently. Technical and ecological solutions appear to be already available, but their adoption by local people appears to depend on new regional policies and education (Anderson 1990a).

Resources and Stakeholders

The actual and potential economic uses of the intact rainforest ecosystem are numerous. The forest can produce a diversity of woods, as well as a variety of edible or medicinal fruits, leaves, bark, game, and fish. The rainforest is the main source of "natural" soil fertility and protects the soils against erosion (Jordan 1987). From an industrial and medical point of view, it is an enormous reserve of molecules and genes, which might prove extremely useful in the future. The rainforest is also a great "carbon sink," which affects the world's climate and atmosphere in largely unknown ways.

The first level of stakeholders consists of primary users: native Indians, farmers, fishermen, traditional extractivists, loggers, and ranchers. Secondary users include the agro-industry, the sawmill and wood industry, the pharmaceutical industry, the tourist industry, and the regional population in general. At the regional and national level, the state is a stakeholder through taxes received and subsidies allocated. In addition, companies that hold land for either productive or speculative reasons are stakeholders. At a third level are the national government with its concern about territorial integration and security, scientists studying the rainforests, and national environmental groups. At the international level, the rainforest is a global concern for the United Nations and for international environmental nongovernment organizations (NGOs).

Indirect stakeholders are the Brazilian citizens who do not reside in Amazonia but can expect to derive various benefits from the maintenance of the forest in the future; for example, through tourism or public income derived from biodiversity. Finally, the world population in general is concerned with climate change and biodiversity losses.

What is the main resource at stake? Each stakeholder has his or her own definition: for the climate expert, the resource is million of tons of carbon; for the entomologist or bioengineer, the resource is biodiversity; for the wood industry, the resource consists of valuable timber trees; and so on. To begin negotiating with these groups, it is important to understand their viewpoints.

Ecological economists have developed methodologies for assigning monetary values to the various uses of a given resource or set of resources. Their indicators of environmental impact can point to which activities are less damaging to the environment compared with others, thereby providing an objective basis for planning. In theory, they can therefore compare the present and future value of different types of managed ecosystems (Kumari 1996). Such methods are useful for policy decisions within top-down planning approaches. They are less relevant when a participatory approach is used. The various stakeholders generally give different values to different uses and are not likely to agree on the economist's choices. In practical terms, elaborating on sophisticated indicators is generally not the best way to start an environmental discussion.

The Setting

The Regional Background

The Transamazônica region extends from Pacajá (220 km east of Altamira) to Ruropolis (340 km west) on the Transamazonian Highway (see Map 2, p. 51). This is the main area of public agricultural colonization, opened by the military government in 1972 by means of public subsidies and government-planned centralized programs. The government had three objectives. First, it was preoccupied with possible foreign claims over Amazonia. To avoid this, the government wanted to increase the Brazilian population in the region. The official slogan was integrar para não entregar (integrate in order not to give away). Second, there was no need to conduct a politically painful agrarian reform to correct the highly unequal land distribution throughout Brazil, since free land could be offered in Amazonia to the landless. Third, there was widespread optimism, fueled by agronomists mainly from the United States who were schooled in the green revolution paradigm, that Amazonia could become a highly productive agricultural region, given the sufficient use of fertilizers and modern technologies.

Before the opening of the road, the region was occupied only by indigenous tribes in the interior (most of whom had no contact with white men) and by sparse traditional caboclo (frontiersmen) populations along the main river, who lived from fishing and rubber extraction. The population grew rapidly during the first decade, owing to government incentives and propaganda (the public slogan was "Amazonia: a land without men for men without land"). However, the number of farmers who actually settled was much lower and the costs per farmer much higher than the very optimistic initial previsions. A large proportion of the farmers, disillusioned with the lack of infrastructure and the low fertility of soils, abandoned their land after a few years but were replaced by newcomers, mostly landless migrants from Northeastern Brazil. After a few years of poor results, the objectives of the settlement program were revised drastically and government support was reduced (Moran 1981; 1989). However, the flux of migrants into the region continued spontaneously during the 1980s and early 1990s and later decreased gradually.

During the early stages of the colonization program, feeder roads were opened and 100-hectare plots were distributed along both sides of the main road in a strip 12 km wide. Feeder roads perpendicular to the main road resulted in a "fish bone" grid of roads. Newcomers had to occupy lands that were beyond the end of the feeder roads. Later, logging companies and local governments extended the feeder roads to an average of 35 km, although some were more than 100 km long. They were, and usually still are, in poor condition and can be used by trucks only during the dry season.

During the 1990s, the wood industry grew rapidly and the sawmills occupied large areas of public land at the end of the existing feeder roads, fueling a new cycle of land speculation. A significant portion of the precious timber (particularly the mahogany [*Swietenia macrophylla*]) has been extracted illegally from the Indian reserves, with dramatic consequences for the surviving indigenous populations. Land tenure conflicts are frequent, but are not so violent as in southern Pará and are generally settled in a more peaceful way.

Today, an estimated 40,000 families live in the Transamazônica region. Family farming continues to be the main form of agriculture in the region, particularly in the western region where perennial crops (coffee, cacao, black pepper) are more developed. In some of the oldest colonized areas, there is a trend toward land concentration and development of cattle ranching. In other older areas, there is a division of existing lots into smaller plots to accommodate the children of the first settlers. Smaller plots (around 10 hectares) are also common around the rural cities.

As a result of these conflicting trends, the region can follow a path of increased large-scale cattle production and deforestation (as in the Paragomina region) or, on the contrary, the reinforcement of a diversified family farming based on perennial crops and intensified agriculture (as in the Bragantina region near Belém).

The decrease in government incentives at the end of the 1970s resulted in strong mobilization and reaction from local dwellers. In particular, there was a strengthening of farmers' organizations. These movements, initially influenced by the Catholic Church, were united in 1991 to create an organization called MPST (Movimento Pela Sobrevivencia da Transamazônica [Movement to Save the Transamazônica Region]). One of the main challenges of this organization is to demonstrate that sustainable family farming can be developed in the region and should be supported by government and agricultural research. This movement also resulted in the founding of various counties (municípios) centered on small cities along the Transamazonian Highway. These counties have allowed local government and elected councils to develop and become part of the region's life.

The Institutional Environment

Although the public sector was weakened by reduced government support, technicians and structures are still present and exert influence principally on credit policy. EMATER (Empresa de Assistência Técnica e Extensão Rural), the state's extension service, and CEPLAC (the Cocoa Extension and Marketing Board) have, combined, approximately 30 technicians working in the Transamazonian region. INCRA (The National Office for Agrarian Reform) officially continues to control the land distribution, but for practical purposes operates only in limited areas with new settlement projects. The research sector in the region was almost extinct in the 1990s after a peak when EMBRAPA (National Agricultural Research Service) staff in Altamira included as many as 16 researchers.

Farmers became quite critical of the research and extension staff, whom they saw as technically useless and authoritarian, imposing "technological packages" that were never well tested in the region. The technicians continued, however, to have a strong influence on the local political life, thanks to their network of contacts in the government administration and their capacity to attract or redirect public or private funding. For example, two former technicians of CEPLAC each managed to be elected mayor in recent years.

History of the Farmers' Organizations in the Transamazônica

Until the end of the 1970s, practically no autonomous farmers' organization existed in the region. Two large cooperatives were created and controlled by INCRA and a farmers' union in Altamira was controlled by large landowners. The process of organizing the farmers was principally the result of the work of the local Catholic Church, which encouraged the creation of religious communities to unite the colonists and also used Sunday meetings to discuss community problems and start community work. The discussion about creating farmers' organizations began in 1979–1980 by some sectors of the Catholic Church linked with the Comissão Pastoral da Terra (Pastoral Land Commission [CPT]), created in 1975 by the CNBB (Brazilian Bishops Conference) to assist the small farmers and landless rural poor.

Two local farmers' unions were formed in 1983 in Uruará and Medicilandia, which participated actively in 1984–1985 in the first spontaneous movement of the Transamazônica farmers. Starting in 1987, the representatives of these groups decided to organize themselves at the regional level to better negotiate with the federal government. In 1988, they carried out a study of the region's social and economic situation, the results of which were presented in 1989 during a meeting in Ruropolis (Monteiro 1996). In 1988, the municípios (counties) of Medicilandia and Uruará were established, and farmers' unions were officially created there. The first leaders of these two unions used their popularity to gain the elections as prefeitos (mayors), aligning themselves to conservative parties and "betraying" the farmers' union.

A new administration of the farmers' union of Altamira was elected in 1987 to represent the small farmers. In 1990 and 1991, various meetings and discussions were organized in the different districts and in the communities. These resulted in two regional meetings in Altamira, with more than 2,000 participants attending each. During the second meeting, in response to decreased government support for the region, there was a one-week sit-in (accampamento). A document called "General project for the development of the Transamazonican region (PGDT)" was developed and presented to the authorities. This project contemplated not only agriculture and land distribution, but also road construction, health, education, energy, credit, and urbanization. Its objective was to propose solutions for the whole region, not only for the farmers' community. One of the slogans of the movement is "Transamazônica: its opening was a mistake, its abandonment would be a crime."

It was also decided at this meeting to institutionalize the informal regional movement and name it Movimento Pela Sobrevivencia da Transamazônica (MPST). The movement was largely dominated by the farmers' unions and groups, but also included other sectors of the regional society, particularly teachers' unions, health agents, and youth and women's organizations (MPST 1993; Hebette 1994).

The MPST was originally formed by 7 farmers' unions, 14 farmers' associations, 3 cooperatives, 5 teachers' unions, and 2 women's groups. It represented an area of approximately 500 km along the Transamazonian

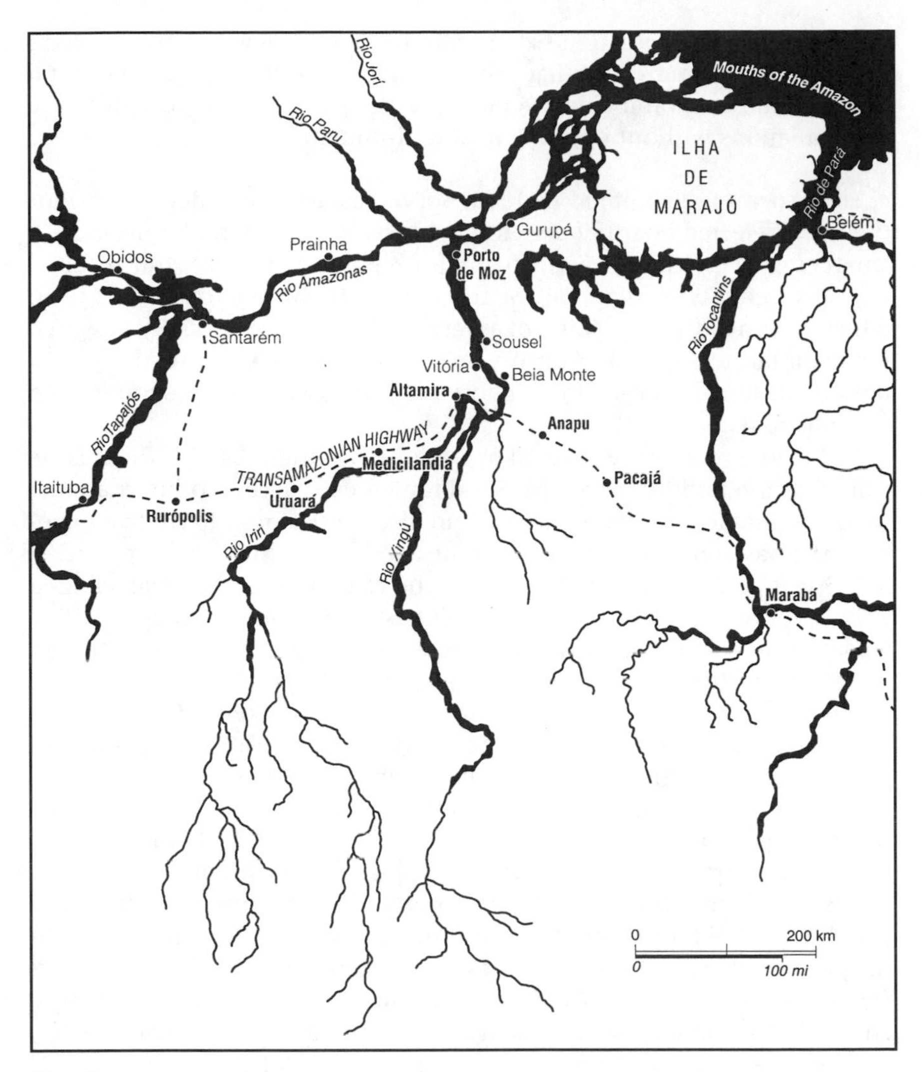

Map 2

Highway (between Ruropolis and Pacajá) and covered a huge area along the Xingu river from Porto de Moz in the north to southern portions of the Altamira district (Map 2).

The general objectives of the MPST (1991) were:

1. To establish conditions that will end the migration to other regions through securing peoples participation in the global development process of Amazonia

2. To organize a debate in the society aiming at new development projects for the Transamazônica that promote human well-being and permit a social and economic improvement for the poorest and most suffering populations without harming the environment

The program, elaborated in 1991, served as a basis of negotiation for the newly elected board of directors with state and federal authorities. A "marching in" with 300 participants took place in Brasilia and resulted in many benefits (conquests) for the region. It was a success for MPST, which obtained the launching of a new credit program specially designed for small farmers (Fundo Constitucional do Norte [FNO]), road repair, a new adult literacy program, support for primary health programs, and other benefits.

The 1991 events were covered by national TV and press. Through fame gained at the national level, MPST obtained easy access to international NGO funding for its own structure and also for various alternative experiences in agroforestry or cottage industries. As a result of the publicity, MPST was able to successfully support the election of local leaders linked with it to local or national responsibilities. One case was that of MPST leader José Geraldo Torres, who was elected as state representative in 1994. Other MPST leaders gained important positions at the state and national levels of the Federation of Farmers' Unions. After 1993, however, MPST structure was obliged to concentrate more on the management of its new responsibilities, particularly in the area of credit distribution, management of small projects, and technical assistance and research, in partnership with a newly created institution called LAET (Laboratorio Agro Ecologico da Transamazônica) with headquarters in Altamira.

With the increasing decentralization of public government planned by the 1988 Brazilian Constitution, and the creation of new districts (municípios) in the region, the power and influence of the elected mayors (prefeitos) grew steadily. The federal and state government policy was to channel most new programs, such as adult literacy and primary health, through these councils. Efforts were made by the MPST to win the local prefeituras (counties), but here they were unsuccessful. The local popular organizations and NGOs, federated in MPST, had to learn to fight with municipal councils and mayors, which remained under the control of the region's elite group of big landowners, merchants, and sawmill owners. Therefore, decentralization, instead of reinforcing democracy, permitted in this case a reinforcement of the local elite and of traditional paternalist power schemes in Amazonia (Geffray 1995). MPST, however, can still be considered one of the strongest and better-known regional farmers' organizations in Brazil. It represents between 5,000 and 10,000 farmers organized in 10 farmers' unions and 23 cooperatives and associations throughout the region. More important than the numbers is the capacity for mobilizing large crowds at given historical moments that MPST has demonstrated.

PAET: Programa Agro-Ecologico da Transamazônica

History

During the preparation of the Altamira acampamento (demonstration) in 1991, the farmers' leaders felt the need for technical and scientific advice that could help to prepare and justify a new proposal for the region's development. They stated, "the research which was conducted by the popular organizations was useful to give support and credibility to this movement by the government, but it could have obtained much better results if it could have been supported by technical assistance" (Medeiros et al. 1995, p. 3). This led to the idea of a structure that could back up popular movements through research in development and help to train the human resources that would be needed to manage new projects and activities.

Some of the local leaders had already contacted and visited another program in the nearby region of Marabá, which was developed jointly by the University of Pará through NAEA (Center for Advanced Amazonian Studies) with four regional farmers' unions. In 1989, they had founded the CAT (Center for Tocantins Agricultural Development), which was based on cooperation between researchers (mostly agronomists) and the farmers' organizations. The foundation of CAT was supported by French international cooperation through the University of Antilles–Guyanne and GRET (Groupe de Recherches et Échanges Technologiques), a French NGO specializing in international cooperation on appropriate technologies and agricultural research development programs.

The MPST then also decided to solicit some researchers from the University of Pará, to support its work and to provide technical advice. Contacts were made with various scholars known for their support of small farmers, some of whom were involved in the CAT project (e.g., Prof. Jean Hebette). In 1992, an opportunity arose to create a local research team with the support of international cooperation (through GRET in this case), with financial support from the European community and the French government. MPST welcomed the involvement, although the participation of foreigners was cause for suspicion by many militants influenced by leftist-nationalist theories. The research team was formed, in

1993, by young graduate Brazilian students, one senior researcher from EMBRAPA, and one French agronomist detached by GRET. It took the name of LAET and signed a three-year agreement with MPST in August 1993. In 1994, LAET and MPST decided to give the name PAET (Agro-Ecological Program for the Transamazônica) to their common activities in the field of development.

The objective of PAET has been defined as follows (LAET 1994):

> To promote the development of family agriculture, sustainable in the long term, and a better management of natural resources, through a research-training-development program, based on a permanent partnership between the farmers' organizations representing family agriculture, and an interdisciplinary and inter-institutional research team.

The main features of the program were:

- To use the participatory action research (PAR) approach
- To be based on a long-term partnership, formalized by a written agreement between a research team (LAET) and a local farmers' organization (MPST)
- To have researchers both in a position of actors (involved in various forms of intervention) and mediators (helping negotiations between the different stakeholders)
- To continually evaluate project objectives and activities and to renegotiate activities as situations change

PAET therefore combined the mediation/conflict resolution methods proposed by Roling (1994), William et al. (1994), other authors previously cited, and the PAR method. This was one of the first times that conflict resolution and PAR had been applied to the field of environmental conflicts in resource management.

LAET's objectives and methods could not be separated from philosophical and ethical choices, which included:

- A commitment to "sustainable development" as a general objective incorporating social, economic, and ecological sustainability
- The possibility of social change through individual interactions (this departs from structuralist views of the human society, in which all behaviors are determined by the social structure)
- Advocating social justice for the poor (justified in part by the idea that a very unequal society is unsustainable in the long term)
- A belief in democracy and education as important tools for sustainable development

An important characteristic of LAET is that it chose to intervene simultaneously at various levels of organization:

1. The first level is that of the *individual farmer*, with the objective of the intervention to improve the sustainability of his or her farming system.
2. The second level is the *community*, with the objective to encourage and enhance common property management.
3. The third level is *municipal and regional*, with the main objective to initiate land use planning within the framework of a regional, sustainable development program.

Propositions of the Program

The principal proposition of PAET was that PAR is an appropriate approach that can both produce scientific results and contribute to solving natural resource management problems. It is a learning process that gives better results in the long term.

There were two subordinate propositions:

- Farmers' organizations are appropriate partners of the PAR/NRM (participatory action research/natural resource management) in the frontier context.
- The platform approach of multiple stakeholders' negotiations on land use planning is an appropriate method of improving natural resource management on the frontier.

These propositions and the assumptions that underlie them are examined in the following chapters as a case study of PAR for natural resource management in the Transamazônica region of Brazil.

Materials

The case study is based on two types of materials:

1. Project documents; minutes of meetings; and reports and papers produced by LAET and MPST and, in some cases, by other individuals or institutions in the region. Some reports and papers are themselves based on field interviews or field work.
2. Personal notes of two kinds:
   a. A report on conversations or debates involving two or more members of PAET, either in formal events (meetings) or in informal mode (personal conversations). As these notes were being taken, there was an effort to analyze the discourses. This approach to formulating hypotheses and conclusions has been termed the "grounded theory" method (Glaser and Strauss 1967).

b. A synthetic description of events produced shortly after the conversation.

Organization of the Case Study

The case study is presented as follows:

- Development of the partnership with the farmers' organization and specific activities undertaken (Chapter 5)
- Three case studies (subordinate to the overall case study) of the municipal participatory planning line of activity using the multiple stakeholders platform method (Chapter 6)
- Presentation and discussion of LAET's results regarding improvement in natural resource management and sustainable development (Chapter 7)
- Analysis of the problems that emerged between LAET and the farmers (Chapter 8)
- Comparison of LAET's diagnosis of natural resource management in the Transamazônica region with conventional views (Chapter 9)
- Evaluation of the participatory action approach (Chapter 10)

PART

THE PARTICIPATORY ACTION RESEARCH EXPERIENCE

The Partnership with Farmers' Organizations

The Starting Point

Partnership with Farmers' Organizations

When Laboratorio Agro Ecologico da Transamazônica (LAET) began its activities in 1993, it chose a regional farmers' organization as a privileged partner for its action research program rather than another type of group. Recent literature suggests that a farmers' organization would be the most appropriate partner for a program aimed at better management of natural resources in the region. For example, agronomists and social scientists working with Farming Systems Research and Development (FSR) have suggested that existing farmers' organizations could best help to define and prioritize the research themes to be conducted (Bellon et al. 1985). Establishment of a partnership with farmers' organizations was also recommended for research and development approaches with objectives broader than technology improvement, such as influencing government policy at regional and national levels (Merril Sands and Collion 1993).

Partnership with farmers' organizations also facilitates the obtaining and collecting of data, and the dissemination of results on a larger scale than is possible with a small research team alone (De Reynal et al. 1995), because farmers' organizations can initiate their own research and extension programs (Fujisaka 1989, Bebbington 1991). In addition, farmers' organizations were believed to play an important role in exerting pressure on governmental organizations, particularly on the research and extension agencies, to make them more client-oriented and to incorporate small farmers' priorities into their agendas. However, examples of effective long-term partnerships between local farmers' organizations and research teams in the field of natural resource management (NRM) are rare, especially in developing countries, and the few exceptions are still at an early stage. An example is the Olafo Project of CATIE (Center for Tropical Agronomic Research and Education) in Central America (Ammour 1994).

Although many bureaucratic and academic barriers to impede cooperation with farmers' organizations were recognized (Collinson 1988), various programs involving a partnership between farmers' organizations and researchers were launched in the late 1980s. A review of the first results of these programs by Bebbington et al. (1994) shows that various problems were encountered in these partnerships:

- The capacity of farmers' representatives to effectively influence research is sometimes limited.
- The farmers' organization seldom represents all farmers' interests. Moreover, these organizations do not always function in a democratic way and the leaders tend to be the first to benefit from the programs (Mongbo and Floquet 1994).
- Farmers' organizations tend to have wide-ranging priorities and draw the researchers into activities for which they are not prepared and are not efficient.
- Most farmers' organizations lack financial and human resources necessary to devote to agricultural research and extension programs.

A notable exception to the last two points are the farmers' organizations or cooperatives that are focused on the production and marketing of one or several commercial products with a lucrative and accessible market. They are more sustainable from an economic point of view and also give more priority to research and development programs that can increase their members' income (Vogel and Krebs 1994).

Assumptions About the Role of Farmers' Organizations in NRM Programs

Despite the problems of early efforts with farmers' organizations, assumptions underlying the programs were still considered valid, and were basic to the LAET (Laboratorio Agro Ecologico da Transmazônica) program.

The first assumption was that farmers' organizations are effectively interested in better natural resource management. This was far from clear at the beginning of the program, but LAET believed that farmers' organizations should be given a chance, since no improvement of resource management could be obtained without the participation of small colonists.

The second assumption was that the farmers' organizations can help in the PAR program by performing the following functions:

- Facilitating the research
- Pressuring the research team for more client-oriented results
- Facilitating the dissemination of results at the regional level
- Representing the farmers in negotiations with the state and other actors
- Facilitating multilevel intervention and the coupling of macrolevel and microlevel interventions

The third assumption was that it was possible to define a common strategy between the farmers' organization and the research team through continuous negotiation and debate. If the farmers' organization and research team had a common interest in improving management of natural resources, it should be possible to reach a common understanding (or a common diagnosis) of the problem to be treated. Based on this common diagnosis and objective, it would not be too difficult to devise a common strategy, and to negotiate the priorities within this strategy. Negotiating a common strategy was therefore seen as the result of a process of progressively improving the communication between the various actors and their mutual understanding of each other's worldview and positions.

The Basic Action Research Cycle

Planning the action research itself was supposed to be evolutionary and cyclical, based on regular evaluations and adjustments in the program (Rhoades and Booth 1982). For LAET, the process was based on an annual cycle, which included the following steps:

1. Elaborating on the researchers' diagnosis (preliminary survey)
2. Expressing farmers' problems and demands (annual regional seminar)
3. Confronting both the farmers' demands and the researchers' diagnosis (annual regional seminar)
4. Selecting priority problems during the seminar or immediately after, with directors of MPST (Movimento Pela Sobrevivencia da Transamazônica), the farmers' organization in Altamira
5. Planning activities to respond to these problems (after the seminar, with MPST directors); activities can be basic research, applied research, training, and technical assistance to the organizations
6. Executing these activities with specific target groups (during the year)

7. Presentating the technical and scientific results to the target group and evaluating the results from the farmers' point of view (before the end of the year)
8. Reviewing the researchers' diagnosis based on new evidence and results
9. Evaluating the progress of the general program with representatives from all organizations that comprise MPST for the next annual seminar
10. Back to step 1. . . .

The basic tenet of this action research method is that all proposed research is negotiated with the farmers' organizations. No research is conducted if the farmers' representatives do not agree with the plan. Research results are then systematically presented (in language understandable to the concerned farmers) and evaluated in common before new activities are conducted.

In theory, the application of action research principles in the planning cycle just presented is straightforward. In practice, however, the application of these principles is not so simple. One of the main questions is about the levels and time frames of discussions. With whom should the research be discussed? Should it be with the MPST representatives at the regional level, with local organizations' leaders, or directly with interested rank-and-file farmers? In the latter case, should we work only with organized farmers, or should we include other individual farmers who do not participate in organizations? Although we agree that research results should be presented to and evaluated by the farmers, there is still the question of when this should be done. Some research results might be discussed immediately after a trial, whereas others need more time to be collected and analyzed. Would a one-year period for general evaluation of the activities be satisfactory? Finally, is it possible to concentrate the discussions only on the research action program, or do we first need to clarify each partner's objectives to define a common long-term strategy? These are some of the questions that LAET and MPST had to face since the launching of their common PAET (Programa Agro-Ecologico da Transamazônica) program. In this section, we will retrace the history of LAET-MPST relations, and later discuss their outcome and some of the lessons learned in this complex experience.

Development of the PAET Program

The discussions between LAET and MPST can be divided into three phases: the first phase (1993) covers the development of a first formal agreement and the definition of first priorities; the second phase (1994–1996) covers the elaboration of a coherent strategy for PAET, with the limited direct participation of MPST; the third phase (1997) is a reaction phase in which MPST openly criticizes LAET and pressures it to reorient its activities. The program that evolved was named the Programa Agro-Ecologico da Transamazônica (PAET).

The First Agreement (1993)

Planning Activities

The three-year agreement that stated the objectives and basis of the LAET-MPST cooperation was signed in August 1993. It stated that MPST's objective was "to propose a new policy for the economic, social and agro-environmental development of the region." The LAET objective was "to assist in the development of sustainable family agriculture (sustainable from the economic, social, and ecological aspects)." The goals centered around the sustainable development of the region and the complementarity of objectives in terms of activities (mobilization, organization, lobbying, and representation of the farmers' interests for MPST; action research, technical assistance, and training for LAET). Both parties agreed to work together as privileged partners, without excluding other agreements with other partners, if such agreements were in line with their respective objectives.

The agreement did not, however, define a common objective and strategy for such cooperation. And it did not define priorities, nor what each partner's specific contribution would be. As a matter of fact, these issues were not discussed in much detail at all between the partners. The agreement was prepared by LAET researchers and handed over to MPST directors. They asked only to modify the description of MPST objectives and background; they did not suggest any other changes. Later developments suggest that the document was seen by MPST directors as a "formality" to satisfy the researchers. For the MPST directors, the important discussion, about whether there should be any cooperation, had already taken place. The idea of a binding agreement between two organizations was certainly new to them. However, both parties recognized that only through practice would the questions of cooperation appear. Regular evaluation was planned internally every two months and annually for the whole program, with large participation from the farmers representing various groups and districts that comprised the MPST. A rediscussion of the agreement itself was not planned, at least not until the third year.

A first regional seminar was organized in August 1993 to present LAET to MPST members, to discuss the first-year priorities, and to make the new agreement official. After presentations of the ideas and trends by the representatives of organizations, and also by technicians and researchers from the public extension and research services, the farmers were asked to form groups and report on their needs and demands for research in three main directions: agricultural production, marketing and processing of products, and the "environmental question." The result was rich, but with a mixture of (a) concrete, down-to-earth proposals, mostly of a technical nature (e.g., how to control the loss of cupuaçu [a fruit related to cacao] seedlings due to predation by rabbits); (b) suggestions for research of particular interest to LAET technicians (e.g., how to improve the use of organic manure); and (c) general wishes, which were sometimes unrealistic or very general (e.g., how to diversify crops and how to get resources to assist in marketing).

LAET analyzed these demands and classified them into five types: basic research, applied research, survey of existing experiences, mobilization of existing knowledge (technical assistance), and mobilization of farmers. Various points that were discarded by LAET at this time (i.e., production of healthy black pepper seedlings and education of young farmers) were later incorporated into its activities. With hindsight, one observes that most activities later undertaken within the PAET were already proposed at this stage. However, they were embedded in a very long list of demands and nobody could predict they would become future priorities.

A few days later, LAET met with three MPST representatives, as well as representatives of the official extension and research institutions, to select its priorities within this universe of demands. It was then decided to prioritize five activities for the following year: (1) a general survey of the regional agriculture and natural resources; (2) a survey of innovative farmers in the region; (3) the follow-up and evaluation of the introduction of animal traction by the Altamira District agricultural services; (4) a study of the agricultural produce marketing at the regional level; and (5) a study of forest wood resources and of the wood industry in Uruará. The first two activities were given priority by researchers in order to construct a better basis of technoscientific knowledge of the region. The latter two activities corresponded to priorities of the MPST administration, which was planning to create a central marketing cooperative at this time. The choice of animal traction was a compromise between an LAET interest in starting work on a technical question that seemed to be of great interest to the farmers and the proposal from the Secretary of Agriculture for the State of Pará to try an animal traction initiative in the Altamira District.

The method of determining priorities by confronting the demands of farmers' organizations with the interests of the researchers illustrates the process that went on during the next years and how it produced a readaptation of the program of activities.

After a few months, the research team felt the need to discuss a more consistent long-term strategy as a result of three perceived problems:

- How to satisfy, at the same time, both the local partners and the project's funding agencies who had agreed to finance a specific project with already defined objectives
- How to face enormous problems, (especially deforestation) at such a scale, with limited resources
- How to combine a traditional goal of agricultural development for farmers' organizations with a proposal of environmental diagnosis that was totally new to all

A first step for LAET was to produce a document describing the proposed program.

Strategic Planning

In August 1994, LAET started a discussion on strategic planning to build a consensus within the team and to form a clearer picture of the future. Initially, inviting MPST leaders was not planned, but at the last moment, one representative was invited. The overall objective was discussed at length until the following definition was formulated and accepted:

> To promote the development of family agriculture, sustainable in the long term, and a better management of natural resources, through a research-training-development program, based on a permanent partnership between the farmers' organizations representing family agriculture, and an interdisciplinary and inter-institutional research team (unpublished proceedings of meeting).

The crucial point for most participants of this meeting was the understanding that the objective of long-term sustainability was incompatible with massive deforestation and that pasture extension had already happened in south Pará after the opening of the Belém-Brasilia highway. Farmers there had initially settled and cleared the land for agriculture, but later converted the land into pastures. Then, as a result of economic or physical coercion, they gave it up to ranchers. The result was a massive concentration of land in the hands of a few, with many of the original farmers becoming landless again. The research team began to realize that, from the farmer's point of view, the joint effort was not directed at protecting the forest or biodiversity for its own sake, but rather at planning a future for local populations.

In November 1994, a second strategic seminar was held, this time with more equal participation of MPST representatives (five farmers, eight researchers). The results of the first seminar were presented and discussed.

The main concerns of the MPST representatives were about the future of LAET (how it could plan to be permanent), its compromise with the MPST political line, and the need for more concrete results in the short term. They asked if LAET could extend itself to other fields of action, such as promoting education, and also to social development in the rural area to encourage the farmers' children to stay on the farm and continue in agriculture. However, few comments were made on the framework prepared by LAET. During the debate, LAET researchers identified various contradictions between the long-term objectives of MPST and its short-term commitments. They also separated the "party politics" of MPST, which could not be supported by LAET as a research team, from a development policy discussion to which it could contribute. As a result, the MPST directors apparently decided to distance themselves from LAET. LAET was not invited to other MPST discussions for several months. Meetings were called during this period only to reply to specific questions or needs.

In December 1994, a regional PAET seminar was organized to evaluate the progress of the program and to develop orientation for the future. It included the participation of 44 farmers from the region. The debate did not bring about very clear directions, possibly because the researchers still dominated the discussion excessively. Much time was spent on technical points of farmers' interest, but little on actually letting them evaluate past activities. The MPST spokesmen noted that the movement had a much wider ambition than to work only on agriculture and that it planned to launch a new mobilization cycle in the region. They also manifested MPST's interest in working with other research organizations in the field of organization, as well as in agronomy.

In June 1995, an internal LAET discussion allowed the preparation of a "strategic planning framework," based on a long-term projection that separated results that could be achieved in the short term (2 or 3 years, corresponding to a project cycle from funding agencies' point of view) from those that could be achieved in several cycles (10 years, more or less) and from those that constituted the long-term objective. The participation of an experienced sociologist (Prof. Jean Hebette) was crucial in helping the team to link sociopolitical objectives (the strengthening of the farmers' organization, the reinforcement of democracy) to sustainable development objectives. Hebette's crucial point was that, in the future, the farmers' organizations should be able to negotiate with local and national governments regarding the implementation of a new model of colonization. This model would be based on intensified and diversified agricultural production systems that concentrated public services in these "intensified agricultural areas." It was hoped that this model would allow a reduction of the occupation of new lands and subsequent deforestation. This framework was then presented to the MPST coordinator for discussion, but with a limited time for MPST to react. The first phase of LAET's program was ending in December 1995, and LAET was already late in sending in a new project proposal.

The Farmer's Criticisms

The MPST representatives were surprised by the extent of this project and the fact that it was based on long-term planning. They questioned whether funding agencies would accept this type of proposal and how the proposal would be interpreted by the farmers. However, they agreed on the general lines of the proposal, but no detailed discussion was conducted. LAET researchers, as a matter of fact, did not allow much chance and time for MPST to discuss the proposal in more detail.

By this time, it was already clear (as admitted by the representatives) that the MPST had difficulty in dealing with the *contradictions* among various frames of reference:

- The need for concrete/short-term results and for construction of a long-term proposal.
- The desire to continue as a large-scale mobilization organization, defending the family farmers' interests at the policy level and, at the same time, properly managing concrete activities (credit, transformation of agricultural products). (The contradiction here was the logic of politics against the logic of economics.)
- The realization that MPST needed some technical and administrative assistance to conduct the various activities, and the fear that technicians could dominate the discussions and that MPST would lose its "producer organization" characteristics.

LAET researchers were convinced at the time that strategic planning had been a very important exercise, effectively permitting the incorporation of MPST views into PAET'S second proposal to its funding agency. However, in retrospect, direct MPST involvement had clearly been limited and restricted to a few representatives. Further, the same representatives rarely appeared at different meetings. MPST's claim that it had not been consulted adequately was therefore correct.

Strategy (1994–1996)

Defining Priorities

Toward the end of 1995 and the beginning of 1996, a new planning exercise was organized for PAET, based on the result of the discussions held at various district levels on the future of family agriculture. Many ideas were proposed and discussed, partly centered around the needs for farmer training, especially in management of farms and of projects (small-scale grants given to farmers' organizations), but also in technical areas. Other proposals were related to an ideal farm designed during the district encounters. These "dream farms" (lote dos sonhos) all had in common an integration of annual and perennial crops, livestock, agroforestry, and fish production. The farmers insisted that research should help them to identify "new crops in which they should invest" to help them achieve this goal. They also asked to be shown new technologies in the field and to have more technical assistance, especially in implementing externally funded projects.

A new need was identified—that of better communication between the leaders of MPST/LAET and the farmers who constituted the base of the MPST. At various times, MPST leaders said that farmers did not understand the role of LAET and knew little of the activities undertaken. LAET researchers, for their part, felt that MPST suffered excessive centralization and that the member farmers did not participate in proposals put forth by MPST, especially those concerning new projects. As a result, both sides agreed to create a new structure to implement better communication within the program.

It was difficult to define clear priorities during this period (1994 to 1996). At an evaluation meeting of a regional seminar on family agriculture, MPST directors expressed conflicting views about the result of the seminar. The general coordinator thought it succeeded well, but another director noted that participant farmers "yet had difficulties to have their own conception of family agriculture" and "they only are able to make proposals on what they can themselves accomplish, but they are not accustomed to work in partnerships." A third director, who was responsible for the farmers' unions, refused to speak but let it be known that he was not satisfied with the process. He apparently thought that LAET was interfering too much in MPST internal questions and priority definitions, and that it would be better to maintain LAET in a more distant position.

In January 1996, the MPST administration revealed the existence of a debate within the organization about its future, with three different lines proposed. One was to maintain the MPST as a mass organization, the second was to transform it into a development-oriented organization, and the third was to change it into a workers' union federation. The structure of MPST was to be rediscussed accordingly. A critical point was the proposal to subordinate associations such as MPST to local farmer's unions; the unions felt they were losing control of decisions that belonged primarily to them. This point was to be discussed during the next General Assembly (to be held in February). LAET realized that the result of this assembly might change MPST's commitment to PAET's program. Nevertheless, the LAET coordinator felt that planning should proceed in spite of the possibility that MPST would place more emphasis on traditional development.

However, during the next General Assembly of MPST in February 1996, no decision was made regarding a change in the structure or functions of MPST. The discussion was postponed until the next year. Another important factor at this time was the election for district council. Two directors of MPST, including the coordinator, had decided to run for prefeito (mayor) with MPST support. This left the administration in the hands of less-experienced members, who were not themselves farmers, but had more of a technical profile and were more inclined to the development line. They decided not to let MPST resources and infrastructures be involved in the election campaign, contrary to the expectations of various candidates and to what has been done during the previous election for state representative with the former MPST leader in 1993. During the rest of 1996, the interactions of LAET with the new MPST administration was marked by an increase in direct cooperation and new collaborative efforts.

The Crisis of 1997

In November 1996, MPST presented its decision to carry out new "internal research" to evaluate its results and strategy, including its partnerships with LAET and other organizations. This evaluation was conducted by a group of agronomists who were external to the region but had the political confidence of MPST for historic reasons. At the same time, serious criticisms were being made about the acting directors, who were thought to be too independent. These criticisms mostly came from some of the historical founders of MPST, now at state and federal positions, who felt that they were losing control of the institution and that the acting directors were too influenced by LAET. They felt that the development line supported by the new administration was going to conflict with their vision of the reinforcement of the popular organizations, and with their own political and electoral objectives. The agronomists also criticized the lack of progress in communication. In fact, there was a basic misunderstanding about what kind of communication was needed. The MPST founders wanted the development of mass communication media, particularly a local FM radio station controlled by the local organizations, whereas LAET was concerned about the lack of personal communication within the MPST, especially between the leaders and the rank-and-file members. This, in turn, raised questions about the internal democracy of MPST.

At the same time, various criticisms that had been raised since 1994 surfaced again. These concerned the PAET program's lack of practical results and the fact that the farm leaders spent more time helping the researchers than the researchers did in helping the farmers. During a new planning effort conducted in the end of 1996, LAET researchers tried to evaluate their contribution to the region's development for the first time. They discovered the necessity of opening a discussion about what they meant by "development" and of comparing their interpretation with that of the farmers. They then realized that the farmers' leaders had a narrow definition of development, which was restricted to increased production, income, improved social facilities, and other things that immediately improved the lives of farmers. In contrast, LAET researchers had a wider definition, which included personal development and the development of a new consciousness concerning the importance of the natural environment for sustainability.

Another finding of the researchers was that the results of the PAET program had been limited for several reasons:

- Excessive dispersion of activities in relation to its staff
- A tendency to give priority to activities and discussions with leaders at the regional or municipal level, rather than at the farmers' and field level
- A tendency to give priority to long-term investments, such as training and discussions on the region's future, rather than conducting activities less ambitious but with more immediate results

These limitations resulted in LAET's loss of some of its credibility with many farmers and with local government technicians, who felt threatened by LAET activities. These technicians seized the opportunity to denounce the competence of the mostly young and inexperienced LAET researchers. As a result, LAET decided to prioritize one or two concrete development activities for the next year, which would have a visible effect and could be publicized at the regional level. Two themes were selected: (1) the establishment of healthy pepper plots in three demonstration units with 15 farmers involved and (2) the experimentation on a larger scale of various types of leguminous cover crops, which had been already tested with some success by some of the farmers. Also, the technical staff of LAET was reinforced to accomplish this effort.

The March 1997 assembly of MPST was rather agitated. The historic leaders confronted the current administration, and took over and reorganized the board of directors. There were also various manifestations of distrust toward LAET. In particular, the renewing of the LAET-MPST agreement (which was overdue) was postponed. During the next months, various discussions about the partnership and new agreement were organized, with various criticisms sounded from both sides. Basically, many MPST leaders felt that LAET was becoming a challenge to them in the region. It had grown steadily and quickly in a way that MPST could not control. However, LAET had not produced the results they had expected and had not reinforced the farmers' organizations.

As a result, MPST made various initiatives to start new development activities without LAET, but with other institutions, and to reduce LAET's participation in ongoing activities. MPST finally proposed that LAET should become integrated in MPST (i.e., under its direction) to fulfill the growing needs in agricultural extension, technical assistance, and research. LAET rejected this proposal. However, by November 1997, both parties agreed that they should restart the cooperation on fresh grounds, after they admitted that both sides made mistakes and that special efforts should be made to avoid such errors in the future. They also agreed to verify that each party saw the legitimate interests of the other side fulfilled in all activities undertaken in common. However, MPST preferred not to sign a written agreement, but rather to rebuild the partnership in a practical way.

Changes in Research Strategy

LAET originally began its project with an understanding that the main impact of human presence on Amazonian ecosystems was occurring during the conversion of forest to pasture or agriculture, whereas other activities such as selective logging and hunting still had more marginal effects (Uhl and Jordan 1984; Johns 1988; Nepstad et al. 1990; Fearnside 1991). Deforestation eliminates the potential for income based on wood products and other extractive products, in addition to depleting soil fertility, increasing erosion, reducing biodiversity, and contributing carbon to the atmosphere. The rate of deforestation was believed to be the best indicator of environmental impact and the main resource of concern was simply forest or, rather, land with forest cover.

The initial project submitted for funding in 1992 was to work on an "environmental impact assessment" of areas that were partly or completely deforested as a result of human activity. The impact was to be based on simple indicators that could give a measure of the sustainability of the production systems in the region (a production system can be composed of cropping systems, livestock and pasture systems, or extractivist systems, separately or in combination). The selection of these indicators was to be made by a multidisciplinary research team. The project also planned to begin a debate with the farmers' organizations in order to establish a common base of discussion on and understanding of their perception of environment and their own long-term future.

The environmental impact assessment was never completed. The LAET team soon came to realize that the farmers had no interest in working on environmental assessment. The farmers' interest was in concrete demands for better living in the short term. The first discussions with the farmers, their organizations, and other local actors quickly demonstrated that environmental planning certainly was not a short-term priority, in spite of the magnitude of the problem. In addition, there was no sign of any long-term planning in the region.

Because rapid depletion of soil fertility following slash-and-burn farming is common in the wet tropics, LAET had initially suspected that this would be a problem for the farmers around Altamira. However, research through questionnaires showed little or no sign of decreasing fertility, at least from the farmers' point of view. Farmers opened new fields because weeds began to choke the old ones. However, they often returned to the original field within a few years. The practice of slash and burn in relatively young secondary vegetation (after 4 or 5 years of fallow) was common, even though forest was still available for conversion. De Reynal et al. (1995) observed in the nearby region of Marabá that the yields obtained after secondary fallow were similar to those after cutting and burning primary forest. The amount of labor needed for each case was also similar, although the seasonal distribution of labor was different. (After a fallow period of secondary vegetation, there was less work for cutting and burning but more work for weeding, compared with clearing primary forest.)

The apparent absence of soil fertility problems can be related to the relatively great amount of land available per family (100 hectares on the average) and to the fact that weeds became a problem before signs of nutrient depletion were apparent. The farmer always had plenty of land, so it was easier to move on and clear new fields than to try to eliminate weeds from an infested plot. However, since a fallow period of 4 or 5 years is insufficient to restore nutrients depleted by several years of cropping (Nye and Greenland 1960), nutrient levels must have decreased somewhat. Eventually, nutrient depletion would become a problem. In fact, Saldariagga (1988) found that several hundred years were required to restore nutrients to levels that exist in undisturbed forest.

Changes in ecosystems did occur. Farmers noted an increase in uncontrolled fires, a reduction in game populations for hunting, a spread of various toxic or hardy weeds in pastures, and finally a reduction of high-quality wood available for building houses or fences. The opinions on deforestation's effect on the changes in local water flow were contradictory. Some farmers observed an increase in the availability of surface water after deforestation; others, a decrease. However, these problems were not cited as the most important for the future of small farmers. Questions of prices, marketing, credit, and technical and social assistance were brought up more frequently. The level and rate of deforestation were seen as private management decisions that each farmer made for his or her own land, even if others criticized the farmer's lack of foresight in foreclosing options that were available through preserving the forest. From the viewpoint of the farmer, cooperating with a team of researchers to study indicators of the impact of deforestation did not respond to the farmer's perceived need and would probably not lead to any fruitful dialogue.

Consequently, the LAET team decided to modify its planning and negotiated a reorientation with the funding agency. This reorientation, justified by signing an agreement with the MPST, moved toward a negotiation of all objectives and activities to meet both parties' interests.

Activities Undertaken

Despite the almost continuous debate regarding directions for the program, concrete activities were initiated and some were completed. Table 5.1 summarizes the set of activities undertaken within PAET and how they changed with time. PAET activities are quite diverse. Some of these activities can be characterized as basic research and surveys, others as collective action research or on-farm research, and others as being more linked with direct technical assistance and communication/training activities. Within each line of activity, the various forms of intervention were used sequentially or simultaneously. The starting points for these activities also were diverse. In some cases, the research team proposed them; in others, they were the result of demands from farmers or local agricultural technicians.

The fact that the program in its entirety can be characterized as participatory action research does not imply that all its activities were "participatory" in nature. For example, in the Porto de Moz activity on natural resource management, a participatory research was first organized, followed by a communication and planning effort (NRM seminar) in which LAET researchers had a facilitator/animator role. Later, there was a training and education component (visits to communities) and then action research was initiated to create four community reserves. At the same time, a master of science dissertation was carried out on the history of natural resource management in the communities.

Table 5.1 shows a tendency for multiplication of activities over time. This is the result of the diverse demands of MPST and the local organizations that constituted it. Many themes were initially suggested by researchers themselves, but in the later years (1995–1997) the trend was to develop activities suggested by MPST. This does not mean that activities suggested by the farmers gave more satisfactory results. As a matter of fact, some activities based on MPST proposals did not achieve the expected results

"(text continues on page 76)"

TABLE 5.1. PAET activities between 1993 and 1997

Activity	Jun 93 to Aug 93	Aug to Dec 93	Jan 94 to Jun 94	Jul 94 to Dec 94	Jan 95 to Jun 95	Jul 95 to Dec 95	Jan to Jun 96	Jul to Dec 96	Jan to Jun 97	Jul to Dec 97	Origin of Proposal
Agronomic survey	X	X	X							X	Research
Sociological survey		I	I	I	x	X			x	x	Research
Marketing of agricultural products	I	X	X	X	x*	x*	x*	x*	x*	x*	MPST
Mechanization/ animal draught		I	X	X	x	X	X	X	X	X	Farmers/local technicians
Agro-ecological zoning (Uruará)	I	X	X	X	X						MPST
Farmers' alternative experiences survey		I	x†		I‡	X					Research
Forestry survey		X	X	X	X		X				Research/MPST
Credit improvement		I	I	I	I	I	X	X	X	X	MPST
Graduate students' training (DAZ)		X	X	X	X	X	X	X	X	X	Research/ university
Pasture management			X§	X§	X§	x§		X	x§	X	Farmers/ technicians
Farmers' management and accounting			I	I	X	X	X	X	X	X	Research
Licenciatura (license) training						I	I	X	X	X	University
CFR (young farmers' school)				I	I	X	x	x	X	X	MPST
Soil studies				I	I				x		Farmers

Black pepper reestablishment	I[§]	X	X		x	X[§]	X[§]	Farmers
Intensification		X					x	MPST
District debates on the future of agriculture			X	X				MPST
Wood processing test			X	x				Both
Natural resources zoning: Porto de Moz				X	X	X	X	Local organization
Communication				I	X			MPST
Introduction of leguminous cover crops					X	X	X	Local technicians
Altamira A-E zoning						X	X	Local organization

I = initial discussion and planning of an activity; X = executing the activity; x = activity continued at a slower pace; DAZ = Desenvolvimento da Agricultura Familiar Amazonica; MPST = Movimento Pela Sobrevivencia da Transamazônica (Altimira);

*Activity taken in charge by German cooperation (Serviço Alemão de Cooperação Tecnica e Social [SACTES]).

†Only animal traction experiences were surveyed.

‡Only agroforestry experiences were surveyed.

§Activity taken in charge by EMBRAPA/CPATU (Empresa Brasileira de Pesquisa Agropecuária/Centro de Pesquisa Agropecuária do Trópico Umido).

and created frustrations. This was the case in the Uruará planning and also in the work on communication. In both cases, failure probably stemmed from different expectations and understandings on both parts.

Some MPST requirements were never attended to, either because they appeared to be out of the scope of the research action proposed by LAET or simply because the team was already committed and did not give a high priority to the demand. In the first category, there were various demands to assist MPST in preparing project proposals for funding. In principle, LAET was open to this type of cooperation. However, some of the farmers' leaders thought that the role of scientists in helping them prepare a project was to listen to what they wanted and to put it into written form that would please the funding agencies. Under these conditions, LAET researchers did not play the game and usually questioned the economic and social viability of the project presented. As a result, they were not called to the discussion again.

Another case was the persistent request for "farmers' training in management," which was brought up in all PAET meetings that had significant farmer participation. MPST leaders, however, never took these requests seriously. There are two possible reasons why MPST leaders felt unable to contribute to this training: first and less likely, they did not want to leave the training to the technicians to avoid increasing technicians' prestige among lay farmers or, second and more likely, they knew about the contradiction between the economic and political logic and did not want to enter into this debate, especially knowing that the majority of actual leaders were better known for their political achievements than for their ability to manage a farm.

In the second category (valid demands that could not be attended), we find requests for organizing more farmers' exchanges and field days, which they found to be a good way of promoting discussions and changes in the farming systems. LAET did not have sufficient resources to organize such activities. Other interesting demands, such as one concerning pasture rehabilitation and management, had to be handed over to the government agency EMBRAPA/CPATU (Empresa Brasileira de Pesquisa Agropecuária/Centro de Pesquisa Agropecuária do Trópico Umido).

In addition, demands from MPST, which were initially rejected by LAET as being out of its scope, were later reincorporated when the researchers realized their validity. The first case is the production of healthy pepper plants. Initially, LAET reacted by saying that it was a classical extension program that should be handed to local extension services (EMATER). However, it became clear that black pepper plants could benefit farmers who lacked the more fertile terra roxa soils (on which more nutrient-demanding crops such as corn and rice could be grown). When EMATER showed an unwillingness to take any responsibility for a black pepper program, probably due to the lack of qualified personnel and excessive burdens with other bureaucratic activities, LAET decided to enter into a demonstration program. With the help of an EMBRAPA specialist, pepper yields were shown to be severely affected by three factors: the transmission of Fusarium disease (wilt) by contaminated but apparently healthy plants, the choice of wrong types of soils for peppers, and the lack of organic manure and mulch for soil preparation (Sakael 1995).

Another example of demands from MPST that were initially rejected is that of rural education for young people. LAET researchers thought that questions of education were outside their scope. However, during a visit to the south of Brazil, LAET invited a representative of MPST to visit various projects conducted by Santa Catarina farmers, organized within a nongovernment organization called CEPAGRO (Centro de Promoção da Agricultura de Grupo), located in Florianopolis/Santa Catarina). Some LAET members thought that the visit would create some interest in the field of production organization by small farmers' associations. However, the MPST representative was much more interested in another experience developed locally, the Casa Familiar Rural (CFR; Rural Family Housing Schools), which proposed a method of alternative professional education for farmers' sons. The young farmers go to the CFR for one week and receive an education based on observation of local practices, analysis, and blending of theory and agricultural practice. For the next two weeks, they go back to their parents' farms and help the family, while they carry out observations and small experiments.

The MPST representative got so enthusiastic about CFR that he immediately started to discuss it with MPST administration and then with others in the Transamazônica region. Further contacts, visits, and training were carried out, to which LAET participated as a facilitator. MPST took responsibility for the process of discussion with farmers and creation of a parents' association, as well as discussion with other government services and county council (prefeitura) for support. The first CFR of Pará State was officially created in September 1995 in Medicilandia. However, the program had to be completely rediscussed because it was originally designed for the completely different agro-ecological environment of the south of Brazil. MPST exerted strong pressures on LAET researchers, who were initially reluctant to assist in this reorganization of the program and also to participate directly in the formation of specific topics. But, as a result of LAET's socioeconomic survey, it realized that the question of children's education was crucial for many families and probably would strongly affect the future of agriculture in the region. Therefore, a social experiment that would permit young farmers to get a minimal primary education, while encouraging them to stay in agriculture, could be of great importance to the sustainability of family farming in the region.

The success of the effort brought home an important lesson to LAET regarding environmental education. Trying to transfer knowledge and sympathy about the functions of nature is not a very efficient way of changing people's attitudes. Helping them to discuss their future in their own terms and within their own organizations is much more useful.

Results of the Partnership with Farmers' Organizations

The discussion of PAET's priorities was a dynamic process effectively involving MPST and representatives of the farmers' organizations. However, priorities were defined in a way that did not always follow the annual planning and evaluation method initially proposed. Some priorities were effectively decided during PAET's planning meetings. However, most came out of response to outside opportunities and others followed a kind of natural selection pattern. Unviable initiatives were abandoned, whereas other activities, originally discarded, were initiated. As a result, it was difficult to set priorities and activities were sometimes dispersed beyond the capability of LAET to carry them out.

The strategic planning efforts between LAET and MPST did not achieve a real consensus, although LAET researchers thought they did. The efforts were certainly important in helping the LAET team discuss and internalize its own coherent strategy, but such efforts did not serve effectively as a discussion forum with MPST leaders, who were neither prepared nor willing to enter into this exercise. The mutual criticisms and the 1997 crisis demonstrated that LAET and MPST did not succeed in establishing a common long-term strategy. In Chapter 8, we discuss the reasons for these difficulties and analyze the various aspects of a relationship established between a farmers' organization and a research team.

Despite the lack of communication at the strategic planning level, a strong mutual influence existed between LAET and MPST. There is no doubt that MPST strongly influenced the LAET program, especially after 1996. The growing pressure from MPST to obtain concrete results in farmers' fields was taken into consideration by LAET in giving priority to "technology transfer" activities (such as improvement of pepper production, previously discussed), which were initially seen as a duty of government extension agents. LAET also affected the MPST program, but in a more complex way. Activities such as the municipal seminars on family farming, the Porto de Moz natural resources survey, and the Altamira agroecological zoning exercise are clearly MPST initiatives that were encouraged by LAET.

6

CHAPTER

Case Studies of the Multiple Stakeholders Platform Method

Assumptions About the Multiple Stakeholders Platform Method Used in Municipal Participatory Planning

Most of the forest in the Transamazon region is still in open-access government lands. The advance of the frontier into such lands occurs through logging and land occupation by farmers and ranchers, often in environmentally destructive ways. Various authors have suggested that land-use planning would be an appropriate measure to control this dynamic (e.g., Fearnside 1986). However, there is a widespread opinion that the usual top-down style of planning will not work and that the local stakeholders, especially farmers, must participate in the planning to give it legitimacy. To ensure that all stakeholders benefit from land-use planning in the Amazonian frontier, there has been an interest in experimental methods of participatory land-use planning. However, practical experiences in this region are rare. LAET's (Laboratorio Agro Ecologico da Transmazonica) opportunity to experiment with the multiple stakeholders platform method in this region was almost unique.

The basic assumption underlying the multiple stakeholders platform method is this: participatory research facilitates communication and understanding between the various stakeholders, helping to create a common image of the problems to be solved, and therefore facilitates conflict resolution and imaginative solution-seeking between the stakeholders.

This assumption embodies three basic principles:

1. It is possible for a group of local stakeholders, competing for different land uses in the frontier, to find solutions for regulating natural resource management that will be in the interest of the majority of stakeholders.
2. Participatory research can facilitate the negotiation process by providing complementary information on the present situation and trends (diagnosis and prognosis) and by helping the various stakeholders to communicate better and to understand each other's interests and views on the future.
3. The state can guarantee that an agreement based on the majority's proposals is reached and respected. This can be accomplished either by compensating the minority losers or by taking fiscal or legal measures restricting free access to natural resources and land.

LAET used the multiple stakeholders platform method for land-use planning in three different districts of the Transamazônica region: in Uruará between 1993 and 1996, in Porto de Moz in 1996–1997, and in Altamira (preliminary activities initiated in 1997). In the cases described here, local organizations suggested, with LAET accepting the idea, that the land-use planning experiment should be included within a broader perspective of municipal development planning.

Case I: Uruará

Why Uruará Was Chosen

A municipality (município) is the basic territorial unit in Brazil. The inhabitants elect a district council, which in turn elects a mayor (prefeito). During the startup of the PAET (Agro-Ecological Program for the Transamazônica) program, a few municipalities had to be selected from the 10 municipalities that comprise the Transamazonian region. After discussions with the MPST (Movimento Pela Sobrevivencia da Transamazônica [Movement to Save the Transamazônica Region]), three municipalities were selected, one of which was Uruará. A decisive factor was the unique dynamics of this municipality since its spontaneous creation in 1978 (Hamelin 1990).

The dynamism of Uruará was demonstrated during PAET's initial contacts with the municipality in October 1993. At that time, a cultural festival had been organized with the support of the Catholic Church and the prefeito. A full day of debates regarding the future of the district was scheduled, and LAET was invited to participate. The main themes were colonization and forestry activity. Several district leaders, including the prefeito, technicians, and Catholic priests, planned a larger meeting with several other institutions. Consequently, the first Conference on Alternative Economic Projects was organized in March 1994. It attracted more than 200 people, representing all categories of social groups from the district and the government, with a majority of participants from the farmers' community.

The Conference on Alternative Economic Projects

One of the principal and very sensitive issues discussed during the conference was that of the Arara Indian reserve, located in the southern part of the municipality of Uruará (Salgado and Castellanet 1997). Lumber companies, with the local support of the Instituto Nacional da Colonização e Reforma Agraria (INCRA), had opened narrow, dirt-feeder roads into the reserve to encourage late-coming farmers without land to settle illegally, thereby giving the companies easier access to the abundance of mahogany (*Swietenia macrophylla*) in the area. Mahogany is the most valued timber in the region; many adventurers have been attracted by this "green gold." The prefeito had proposed new borders for the reserve, excluding the areas already occupied by the farmers, but FUNAI (National Foundation for the Indians [Fundação Nacional do Indio]) and CIMI (Centro Indigeniste Missionario), a regional nongovernment organization for Indians' rights, rejected the proposal.

The conference was an opportunity to try to solve the conflict peacefully by bringing together representatives from the settlers, the prefeito's office, the Ministry of Justice, the Catholic Church, and scientists. CIMI was also invited as an observer. Although a solution was not found right away, this initiative probably prevented the conflict from turning into violence, something that is common in Pará, as occurred during an event in which 77 Tembé Indians were taken hostage by posseiros (land settlers who claim land but have no title), who illegally occupied their reserve in May 1996.

A report on forestry presented by LAET was also the subject of much controversy. It reported that certain farmers were criticizing logging companies (madereiros) for purchasing trees at ridiculously low prices and not always being true to their word. Other farmers, however, noted that the madereiros were indispensable because they were the only ones who opened and maintained the feeder roads, something that INCRA and the prefeito were incapable of doing.

During a plenary debate on this and other forestry issues, representatives from farmers' organizations associated with the MPST had difficulty getting their turn to speak, since the discussions were strongly dominated by the technicians. However, the LAET presentation encouraged many farmers and young urban workers to take part in the discussions. The careful preparation and clear explanation of this presentation, with simple drawings illustrating important issues, certainly contributed to the debates. Several farmers declared later that, for the first time, they had understood what the technicians were talking about, in contrast to most other conferences, which tended to be too technical.

LAET had hoped that the notoriety and social standing of the LAET scientists would have enabled some of the socially weaker participants at the conference to express their point of view publicly. Wherever there is strong pressure for censorship, the ability to tell the truth can bring about change, even temporarily, among the local power relationships. However, many people later confessed to LAET scientists that they were afraid to speak in front of the municipality's "big wigs." One cannot blame them, even if relationships in Uruará appear to be slightly more civilized than in other regions, in a climate in which it is not uncommon to assassinate menacing competitors.

The conference was marked by an "accidental" visit from a candidate in the upcoming elections for governor of the Pará State. This visit demonstrated that the objective of the conference was to address not only the issue of participatory projects, but also the municipality's plan to become a "pilot district" of Pará and thus benefit from the future government's backing.

When the conference came to a close, the most important proposals concerning the forestry issue were:

- The creation of a natural reserve and a pilot municipal forest (to be managed sustainably by the community)
- The enhancement of the value of wood by creating cabinet-making workshops
- Sale of trees at their estimated volume value, not at a unit price set by the madereiros (logging companies).

The grassroots farmers had an interest in studying these alternatives for maximizing and enhancing the value of wood and in creating their own cooperative sawmills. Clearly, this would not be easy to manage. Even the farmers' union of Uruará did not foresee how it could be done. Several times during the discussions, union representatives declared that they were no match against the large forestry companies because of the economic power of the companies.

After the conference, a group of representatives from local institutions at the district level created the Fundação para o Desenvolvimento do Municipio de Uruará (FUNDASUR) to implement these proposals. The district council obtained the backing of SUDAM (Superintendencia do Desenvolvimento da Amazonia), a regional public structure based in Belém, charged with bolstering public investments in Amazonia. They were to conduct a detailed agro-ecological zoning of the municipality through satellite images. The resulting maps were to officially determine the best location for the proposed municipal forest.

Another result of the conference was that LAET drew up a proposal for a second phase to their research on forestry activity, which was presented to the MPST and the Uruará farmers' union. The MPST was interested in this because the urban society in the region—young students in particular—considered logging to be predatory and not very profitable for the region. As a nongovernment organization whose motto is "live, produce and preserve," MPST was thus obliged to back ecological research, knowing full well that its main body of farmers was not exactly enthusiastic about it. The research project would allow LAET to analyze and learn more about the forestry activity in the other municipalities and to enable LAET to draw up a proposal for the management of natural resources on a regional level.

The Study on Forestry Activity

The study on forestry activity took place from October 1994 to February 1995. Its objective was to learn more about forestry activity in other parts of the municipality, to establish measures for the local public management of forests. The study included evaluating the forestry activity, the actors and various marketing possibilities, and the costs and revenues for each phase. Semistructured interviews were conducted with sawmill owners, as well as forestry and sawmill workers. Further information was gathered from informal conversations with local inhabitants. Each phase of the research project was to be discussed with the union so that the information provided would raise interest and reinforce the farmers' organizations.

Results

Regional Background

Before agricultural colonization, the central region of Pará, where the humid tropical forest constitutes most of the vegetation, was known for extraction activity, particularly that of Brazil nuts and rubber. A large part of the region near the Iriri river was marked as Indian reserves. Logging and lumbering were restricted to the forests of varzeas (floodplains) along the Xingu river, where only one sawmill existed. The terra firme (upland) forests became accessible to logging only with the construction of the Transamazonian Highway.

In the 1970s, the Uruará region had only two sawmills. In the 1980s, three large companies began mahogany extraction in the Iriri river region. In 1993, Uruará had eight medium or large sawmills, three of which belong to large export companies. There were also approximately a dozen smaller sawmills in the countryside and about 50 small loggers—intermediaries who did not have sawmills (Salgado 1995). By 1994, logging represented 45% of the municipality's estimated primary revenues.

The Timber Resources

Forest inventories in the Transamazonian region showed a wide variety of wood and significant volumes for commercial use (Dantas and Muller 1979; IBDF 1975). However, only two or three types of high-value wood (*Swietenia macrophylla*, *Cedrela odorata*, and *Tabebuia* spp.) were extracted for commercial purposes during the early logging activity of the 1970s. After these wood species were exhausted and sawmills using these species dwindled, approximately 15 other types of wood of a lesser value were extracted.

To estimate the volume of wood available in the Uruará municipality, the surface area of the municipality's forest was multiplied by the volume of marketable wood, which is estimated at 6 m^3 per hectare according to the IBDF (Instituto Brasileiro Desenvolvimento Florestal [Brazilian Institute for the Development of Forestry]) inventory. The volume of wood extracted since the opening of the Transamazonian Highway was then deducted, using available statistics. As of 1997, an estimated 3.5 million m^3 of wood, of presently commercial species, remained to be extracted. At that rate, this means 30 years of supply before all commercial timber is exhausted.

Uruará has approximately 3750 farming families. Because each family deforests an average of 2.5 hectares/year per farmer, 56,000 m^3 of marketable wood is cut annually. This potential, therefore, represents half the present production of the municipality. Unfortunately, fallen logs from areas cleared for farming are only partly used by the industry, owing to the low number of logs on each property and the difficulties in collecting such scattered resources.

Logging Procedures

Large sawmills generally boast very large teams and equipment including several bulldozers, trucks, and other equipment. Smaller mills do not always have extraction teams and most of the logs are bought from the toreiros (those who transport the wood in logs on small trucks). Their trucks and tractors are usually bought on credit (reimbursable in wood) from the sawmill owners. Most toreiros were, or still are, farmers. Logging is their principal activity during the dry season; during the rainy season they dedicate themselves more intensely to farming.

The loggers' working radius depends on their available capital. The toreiros cover a range of up to 50 km from the sawmills, teams from average sawmills cover a distance of up to about 100 km, and those from large sawmills up to 300 km. As the quantity of valuable wood decreases, these teams move on to other regions rich in export wood.

At the time of the study, an estimated 50% of the overall volume of wood extracted from the municipality came from private lands and the other half from government land and Indian reserves. According to law, all marketed timber should come from managed forests certified by the federal environment agency, Instituto Brasileiro do Meio Ambiente (IBAMA). In actuality, forest management was practically nonexistent in the region, with the exception of limited "showcase" areas of reforestation set up by big sawmills.

Timber was bought either per individual tree from the farmers' properties or in the form of production rights (i.e., a percentage of the final value of the volume produced) on large properties (fazendas). However, it also could simply be pillaged in the Indian reserves and on government lands. In Uruará, for instance, two 400-km^2 properties of approximately 40,000 hectares each were bordered by two large export companies in 1993–1994. According to a declaration from the president of the Institute for the lands of Pará (ITERPA, Instituto de Terras do Pará), it was later discovered that the corresponding land titles had no legal value. These lands have, nevertheless, already undergone logging operations and their forest management plans were accepted by IBAMA. Often, these same companies later set up pastures. These phenomena can be observed throughout the region. In some cases small farmers have been expelled from their lands by large companies.

A great portion of the wood production of the large companies is not sawed in the region, but transported by river to Belém. The estimated total production of the municipality in 1994 was 115,000 m^3 in logs (Salgado 1995). Sixty-five percent of this volume is probably handled by large companies, 22% by average sawmills, 12% by small sawmills, and 1% by farmers who use only a chainsaw. Most of the production from large sawmills is sold for export to international trading companies. A large portion of the production from medium-sized sawmills is sold either to export companies or to the domestic market, particularly to the south and southeast of Brazil. The production from small sawmills is sold essentially to tradesmen from other cities, who come to buy the wood directly from the mill, or to the local market.

Economic Aspects of the Wood Industry

The prices paid for standing trees in farmers' lots vary according to the relationships established between loggers and farmers. Some farming groups negotiate with loggers and give them the trees from all the properties located along an access track in exchange for the opening of a dirt road, which is indispensable for selling their farm products. Others negotiate individually, but the prices are often reduced as a result of the debt that most toreiros have with sawmill owners.

The price for an individual tree in 1994 varied between 12 and 30 reais (depending on the species). The price per cubic meter of logs at the sawmill fluctuated between 20 and 60 reais, and the price per sawed cubic meter (produced from the sawmill) varied from 100 and 337 reais, depending on the type of timber (1 real was equal to $1 US at that time).

An average sawmill in the region processed 5000 m^3 of wood per year, and the average distance that each log was hauled was 60 km. The average mill achieved a 40% processing efficiency (100 m^3 of raw log resulting in 40 m^3 of sawed boards). Based on the value of cedar (*Cedrela odorata*), the type of wood most extracted in the municipality in 1994, LAET estimated that the net margin for the industry was 45%, whereas the return to farmers was only 5% and to industry workers 8%. As a result, the local value added in the município economy was small. Most of the money never reached the município.

In small-scale operations in which timber was sawed with a chainsaw, profits were lower, but they were more labor-intensive and less capital-intensive. As a result, most of the income stayed locally with the farmers, who could multiply by a factor of 10 the income obtained from timber. They could also work at their own pace, without depending on heavy equipment from large sawmills.

A low price per tree is generally considered to be one of the causes of the poor management of forest resources. Encouraging farmers to increase the selling price of individual trees should ease the pressure and lead to a more rational management of the forest. In this case study, however, in which a large proportion of the wood extracted by large companies is located on government land or on Indian reserves, it is unlikely that such an increase would slow logging activity down.

Environmental and Socioeconomic Impacts on the District

The selective logging of certain species of trees, such as mahogany and cedar, has resulted in the exhaustion of those of marketable size. These species can now be found only at great distances from the sawmills. Marketable cedar can be found 80 to 100 km from the sawmills, whereas mahogany is found only at a 300-km distance. However, the direct impact on the whole ecosystem is probably limited, because densities of exploited trees are low and canopy is quickly restored (Verissimo et al. 1992).

The most negative impacts arise more from the *consequences* of logging than from the logging process itself. The routes opened in the forest are farther and farther from the Transamazonian Highway. These openings are soon occupied by families who sometimes already own the lots and who clear the area for crops or pastures (Castellanet et al. 1998). This leads to even greater deforestation.

Many socioeconomic problems arise from the disorderly occupation of space. Maintenance of the dirt roads has become difficult, if not impossible, for the prefeito because the roads sometimes extend up to 90 km from the Transamazonian Highway. It is not easy to transport farm products all the way to the main road. Due to high transport costs for products, profit for the farmers who live along these feeder roads is low. Furthermore, transportation problems make it difficult to market fresh products. As a result, farmers who settle along the feeder roads cannot obtain a reasonable income from farming. Their access to social services such as schools and medical clinics is also severely limited. In addition, isolation makes it difficult for their associations to meet regularly to discuss long-term projects. Added to this is the problem of fire in areas that have been logged recently. Logging takes place during the dry season and leaves many clearings with dry wood on the forest floor. This makes the area a prime target for fires, which often spread into fields and destroy perennial crops.

Another important negative impact is the loggers' invasion of Indian reserves. In the Brazilian forest, mahogany trees are located on 79 Indian reserves. In the Transamazonian region, as well as along the Xingu and Iriri rivers, most of the reserves have been invaded (CEDI 1993). The social impacts are very strong when there is direct contact between the madereiros and the Indians. Contacts with the "white men" frequently result in destruction of Indian tribes, especially due to alcoholism. Indians gain cash for alcohol by selling trees, often at prices lower than those offered to farmers. Selective extraction also causes significant environmental impacts on the vegetation, particularly on game, which is the staple food of Indian tribes. The occupation of the reserves by farmers also is a serious concern. Currently, 350 families occupy the Arara reserve in Uruará.

The Municipal Seminar on Forest and Wood

LAET's study on forestry activity was presented during a March 1995 Forestry Development Seminar. The seminar brought together madereiros and representatives from associations and communities, as well as local and governmental institutions such as EMBRAPA (Empresa Brasileira de

Pesquisa Agropecuaria, the Brazilian Agency for Agricultural Research) and SUDAM. The results of the study were strongly criticized by logging companies. They did not agree with the estimation of their level of profits, nor with the evaluation of the total volume of wood extracted from the município, since it was twice that indicated in the official statistics. LAET's estimate indicated a serious level of tax evasion on the part of logging companies. Although the farmers' union had encouraged the study, it hardly participated in the seminar at all; its representatives apparently had mixed opinions about the political nature of the seminar.

The most significant resolutions adopted at the close of the seminar were:

- The creation of a cooperative to produce and market the wood
- The collection of waste from sawmills for cabinet-making workshops and alternative forms of energy
- The cultivation of forest plant species
- Training for farmers in various fields of forestry
- The construction of a port route (up to the nearest navigable river) to reduce the cost of transportation and facilitate the selling of the wood

Proposals for Enhancing the Value of Wood for Small Farmers

The Forestry Development Seminar aroused the farmers' interest in the idea that they could sell their wood at a much better price if they could process it themselves. They were also interested in sawing the boards themselves to use in building their own houses and furniture. LAET's research suggested that, as a first step, sawing logs by chainsaw might be an appropriate technology. It could enable farmers to saw and market the wood without making large investments. Transportation costs of sawed wood would be lower than costs for logs. Farmers could work at their own pace and stock the sawed planks for a few months, until the roads were repaired. They could also make use of logs cut during slash-and-burn agriculture and then abandoned, either because the logs are too dispersed or because the trucks cannot reach them before the end of the dry season, when the wood on the fields is burned.

The creation of a cooperative was the subject of vigorous debate. The problem was that sawing logs with a medium-sized mill and then selling the boards would probably give rise to management problems. Such an enterprise would require the management of an industrial structure, which is tedious and difficult to maintain. LAET recommended leaving this idea for a second stage.

In October 1995, the Sindicato de Trabalhadores Rurais (STR), the rural workers' union, invited 15 farmers to participated in a four-day course in Uruará, about the better use of wood resources. The object was to show them how to estimate the potential value of their forests, and to study how to saw the logs in situ, evaluate the costs, and compare them with other possibilities (sawmill, sale of wood in logs). This seminar brought together farmers and scientists, and appeared to be fundamental for putting ideas into practice. It was a time of greater interaction between farmers and researchers. During the course, the farmers participated in calculating the number of trees on their land and their cubic volume, in estimating their value and losses due to sawing of the wood, and all other calculations.

The course encouraged the initiation of a collective experiment for maximizing the use of wood. Included was a practical exercise: sawing an abandoned cedro (cedar) log in a farmer's field. Then, calculations on the costs and benefits of this activity were made with the group. This confirmed the high return of the activity, after a market for sawed boards was established locally. One of the sawmill owners of Uruará said that he would be interested in helping to market the boards. However, it was difficult to verify whether he was sincere or whether his strategy was to gain the farmers to his side.

Although most of the farmers in the course had difficulties with the calculations, they were finally convinced that log-sawing could be a highly profitable activity. They also saw, in practice, how one could conduct a survey of its forest and then evaluate the standing volumes and their value. They even discussed whether they could enrich their forest in valuable timber species through direct plantings. However, some of the leaders were still resistant to the idea of simply using chainsaws to cut boards, since they thought that it was a regression in technology. They had been expecting proposals for bigger equipment, such as a mobile sawmill that could serve various farmers sequentially and give more prestige to the Farmers' Union.

It was finally decided that several tests would be started with two or three communities selected by the STR. LAET was asked to produce an easy-to-read document that could be distributed to farmers so that they could remember and publicize the results of the training and practical exercise. However, the process dragged on for some time. A few weeks later, a representative of the STR informed LAET that the farmers had re-discussed the proposal and found that using the chainsaw was not a good solution. It would be too tiring and dangerous.

One of LAET's biggest limitations in Uruará was that no team member actually lived in the district (Uruará is 180 km from Altamira, the equivalent of about 6 hours of driving and, until 1997, had only one public phone). LAET lacked follow-up. The most active researcher in the forest management left in the beginning of 1996 and things continued undefined for a few months.

Meanwhile, LAET was contacted by consultants for the pilot project for Amazonia (PPG7) described in Chapter 1. They were looking for promising projects in the field of forest management and had heard about the Uruará experience from several sources. After a rapid visit to Uruará, the consultants recommended that the project for enhancing the value of wood should be given to FUNDASUR for implementation. This decision created a conflict between LAET and the Farmers' Union, since both groups had already agreed to work together on this project. However, the project proposed by FUNDASUR was more ambitious and bigger than that envisioned by LAET. FUNDASUR planned to directly organize forest management on a cooperative basis by joining together all the forest plots from different areas and then forming a collective forest that would be managed sustainably.

As a result, nothing happened for almost a year. All involved were waiting for the FUNDASUR project to be approved. Then, in October 1997, the local press announced that the project had been approved by the funding agency, but the money was not yet available. In November 1997, the MPST board of directors became impatient and decided to reopen discussions with LAET to try to get things moving. However, as of November 2000, the initiative for forest management in Uruará is still waiting to be launched.

The effort to give the forestry management project to FUNDASUR demonstrates very well the negative effects of inappropriate funding on the local dynamic. The idea of PPG7 to support "promising initiatives" in the area of forest management was certainly based on good intentions. However, the project philosophy was still the same as those previously adopted by banks and international funding agencies, that of a "blueprint approach." Each initiative was to be selected on the basis of expert analysis, to have a limited duration, and a defined scope and planning. Due to these restrictions, the funding agencies (including the World Bank) preferred to fund a few big projects of limited duration and to discourage small-scale initiatives based on locally available resources. This is in contrast to participatory action research (PAR) projects, which by nature must start small and grow progressively, based on experience gained, and must have long-term perspective and duration (to allow periodic readjustments based on regular evaluations of progress). Supporting a learning process rather than a blueprint approach for rural development is the more appropriate form of aid and offers better results in both cost/benefit ratios and intangibles such as conservation of biodiversity (Korten 1980).

The flexibility of funding agencies is critical. The basic idea of LAET was to start small and to carry out tests with limited external inputs, so that the farmers themselves and their organizations could learn by doing and evaluate the possibilities of scaling up on a realistic basis. However, the temptation of big pots of money was hard to resist. All the local organizations, from the STR to FUNDASUR, the district council, and even MPST, found it more interesting politically to obtain a big project funded from outside, which would bring in brand-new equipment and easy money.

Conversations with individual farmers suggested that many were genuinely interested in using chainsaws to saw logs, once they could sell the end product. But they were in fact discouraged by technicians or by their own organization (STR) on the grounds that this was a primitive technology. The conclusion is that when a funding agency lets it be known that a big project might be financed, increased competition between the local organizations results and, worse, efforts underway at the ground level are discouraged. To avoid or at least to reduce these risks, researchers involved in local projects should maintain direct contact with the farmers as much as possible. But this means time and dedication in the field.

Other Proposals

During the municipal seminar, proposals for the creation of a municipal forest reserve, as well as a local tax and environmental controls office, were voted in by the madereiros, even though the proposals were not favorable for them. Perhaps they believed that these decisions were just a façade and would never be truly implemented. Perhaps they believed the gains from agreeing to create a municipal forest would largely compensate for the losses incurred from the other measures. The gain they hoped for would be the opening of a route to a river port in the north of the municipality. The loggers needed a road, but the State Secretariat clearly indicated that it would be difficult for the government to back the creation of a new route due to the obvious environmental risks in this type of investment. A good strategy for the loggers, therefore, would be to exchange the route for a municipal reserve, which would end up occupying a maximum of only about 10% of the territory.

Another proposal concerned the dealings between sawmill owners and farmers. Some owners affirmed that, in some cases, they preferred to negotiate the extraction of wood with local farmers' groups. When they negotiate with each settler individually, they risk not being able to purchase the wood from certain lots, even though they must open paths throughout the territory. Farmers who refuse to sell after the paths have been opened later benefit from selling their trees at a better price to small competitor madereiros, who have not paid anything for opening the path. If all the farmers agreed to sell simultaneously, the logging could be organized in a more rational manner and would result in reduced costs. In theory, the profits from this improved organization could be divided between the settlers and the madereiros.

Some proposals favored the economic interests of both the settlers and the timber companies. For example, foresters gladly back settlers in reforestation projects on the farmers' cultivated lands. After all, it doesn't cost the foresters much, and it tempers the public's criticisms regarding deforestation. It may also ensure their future in the long run. Likewise, the possibility of producing energy at a lower cost (and more regularly) from ligneous waste is in the interest of everyone, including farmers who have a part of their family in the city of Uruará. Sawmills incur heavy costs in producing their own energy through diesel generators, because the local company cannot provide electricity on a regular basis.

There was not always agreement among the forestry companies themselves, especially regarding their relationship with farmers. Small, family-run sawmills, a part of the municipality for a long time, did not appreciate the arrival of large mobile companies specialized in rapid and brutal wood extraction far from the principal routes, which then left after the damage was done. Small sawmills are more concerned with maintaining good relationships with small farmers, who are their principal suppliers.

Impact of the Participatory Planning—Two Years Later

By 1997, the results of participatory planning in Uruará were disappointing. The municipal forest proposal did not go forward. On the contrary, the area that was proposed by the ex-prefeito in the northern part of the district, had been extensively exploited by sawmills who opened their own road to the Curuá river northward, even though they did not succeed in getting public support for this project. Officially, the new prefeito was awaiting the analysis of remote sensing images by SUDAM experts to determine the best location for the reserve. However, all indications are that the proposal will be forgotten.

Farmers also did not even try to organize themselves into a cooperative to process and market wood. The local representatives believed that the first step should be a project to attract funding and obtain heavy equipment; only then would it be necessary to discuss the cooperative. Initial efforts to organize processing at the local level did not go forward as a result of the union leaders' preference for large-scale projects and the selection of the district as a pilot site under the PPG7 program.

On the other hand, some impact was observed at the community level. After the seminar, certain communities decided that they would no longer sell logs at a low price to sawmill owners. Others were more firm in imposing contractual agreements to loggers (in particular, stipulating that they repair and improve the road after logging). Although the relationship between farmers and loggers did not change from an economic standpoint, there was an improvement in the fairness of transactions. Sawmill owners agreed to these changes because they realized it was in their interest to improve the image of the wood industry.

Indirect and Institutional Effects

The Uruará conferences were an opportunity to invite representatives of various public services and the government. The participation of these representatives focused public attention on the district and resulted in further benefits to the município. For example, Uruará was the first in the region to obtain a new agricultural credit line oriented toward family farmers. It also permitted the establishment of a subsection of INCRA (Agrarian Reform Office) in the district and a new farmers' settlement project (Campo Verde).

The decision to prioritize Uruará was part of a scheme worked out by the newly elected state governor, together with the local priest and other local political figures, to gain increased support for the Transamazônica region. As the governor said in a meeting with regional representatives on August 3, 1995: "We cannot spread our funds over all the Transamazônican region. But we may support Uruará as an example of how people and institutions should work together for their future." In this respect, the final conference, which prepared a general development plan for the district and was held shortly before the next prefeito election, was more of a political mark than a technical one. As a matter of fact, the plan itself was more of a shopping list than a real municipal program with clear-cut priorities and explicit trade-offs.

These efforts did not produce the expected results and the governor's candidates were not elected, despite the important economic support from the sawmill owners, who repaired various feeder roads free of charge during the campaign. Instead, another ex-union president was reelected, despite serious corruption charges made against him. The traditional "clientelist" and "paternalist" attitudes were still very much present in the frontier countryside.

The defeat of the governor's candidate, who was an ex-president of the Farmers' Union and an MPST coordinator, provoked an internal crisis in the MPST. The remaining directors did not accept the ex-coordinator's return to his seat after this disappointing defeat. Also, various criticisms were made internally against the way he had conducted the alliance with the governor, especially because it left the Farmers' Union isolated and weaker than before. Also, various members of the Uruará Farmers' Union did not vote for him—in a clear attitude of disagreement against this strategy.

LAET, for its part, has been marginalized from the Uruará process, especially after the wood seminar. More and more, LAET was not invited to meetings. Its own "restitutions" or training sessions in Uruará were dominated by local technicians who gave the credit for the work to the FUNDASUR foundation. Some of LAET's work directly with farmers was undermined by at least one influential local agricultural technician. But there was no serious discussion with MPST about these occurrences. A powerful anti-LAET group had been formed and was joined by the sawmill owners, who saw their direct interests threatened. The technicians saw LAET as a dangerous challenge, especially one local representative who knew that some of his past activities in the district could be criticized. The technicians may also have made an agreement with a local religious leader who felt that his paternalistic methods were being challenged by LAET's criticisms. As a result, the officials of the Farmers' Union and MPST agreed to "keep LAET more distant from Uruará."

Lessons from the Uruará Case Study

The underlying and implicit hypothesis of the Uruará project was that better knowledge and explanations of the perspectives of various resource users would help in the ongoing negotiations between the stakeholders. This approach would be useful in finding solutions acceptable to everyone, or at least to most users, in the interest of the majority and of future generations.

Although a number of innovative proposals were formulated during the forestry seminar, it later became evident that these proposals had been negotiated between different influential local groups and did not generally ensue from a discussion of the study's results. Furthermore, LAET was marginalized by the local elite at the crucial time when proposals should have been transformed into either projects, training, or testing with the farmers. The researchers concluded that they had been used by local leaders to legitimize these proposals through the state representatives, who were, in fact, the main target of the conference. It was believed that these representatives could mobilize new projects and funding for the district. LAET was also involved unwillingly in a large-scale political scheme that was not very conducive to real participatory discussion and ground-level development effort. They finally concluded that the most powerful user groups would end up benefitting the most from this project, either directly from the resources that could be mobilized or indirectly through enhancement of their political image.

The major lesson of the Uruará experience was that the dynamics of local planning could not be understood without also analyzing the objectives and strategies of the numerous groups that did not use these natural resources directly: technicians, politicians, tradesmen, churchmen, professors at the local level, and public and private organizations at regional and national levels. LAET suffered from the lack of experienced members (in the local political context) and the lack of a continuously present researcher or technician in the Uruará município. Without this continued presence and direct contact with farmers, a personal, confidential relationship could not be developed with locally well-informed actors, who could have informed LAET about some of the local strategies of the different groups. Today, it is evident that the effective participation of researchers in participatory planning requires a good study and analysis of the local sociopolitical situation. Without a minimum understanding, it is difficult to distinguish between rhetoric and sincere declarations. Researchers must establish an action plan to avoid being manipulated by well-organized local interest groups to the detriment of real democratic participation of the most numerous, but less influential, groups in the society.

Another important lesson was that it was unrealistic to expect that, by simply presenting the results of a preliminary "stakeholders' analysis," a discussion could be held on an equal basis among the different participants, leading to a conclusion in which the general interest would prevail. In the rhetoric of the seminars, this common interest was unanimously set as the objective. The Catholic priest played a particular role in this rhetoric by declaring that what was good for the rich was also good for the poor and that everybody should unite in the municipality's interest. This type of discourse systematically negated the fact that conflicts of interest existed among various groups.

The consensus that all proposals should be adopted unanimously resulted in each seminar or conference ending with a long list of proposals and projects. The local elite knew perfectly well which of these projects would be quickly forgotten and which would be actively supported by members of either the FUNDASUR or the prefeitura. By eliminating the debate and conflict from the public scene, all the real choices were left in the hands of the powerful, who later used their influence to support one or another project.

To establish the balance in favor of the numerous, powerless actors, it seems necessary to concentrate on helping these groups to reinforce their own organizations and to develop their own proposals before they reach a negotiation table. These are people who are economically weak and do not have access to the technical and administrative knowledge needed to interact with government experts and technicians. Similarly, Fals-Borda and Rahman (1991) concluded that the situation of the less-powerful groups can be improved only if researchers establish a strong initial alliance with these groups and if a forthright, unambiguous, common strategy is discussed with them, rather than trying to deal with all local groups and actors equally. These lessons were taken into consideration in the next case study, that of Porto de Moz.

Case II: Porto de Moz

Porto de Moz is a district with a more traditional population than that of Uruará. Located along one of the main rivers, its population is predominantly caboclo, people of mixed race who have lived near the Amazonian riverbanks for many generations. Catholicism is the predominant religion, especially in the rural areas, contrary to other situations in the Transamazônica. The main economic activity was traditional extractivism until the 1960s, but logging has played an ever-growing role in the local economy. Today, wood extraction (mostly in logs) represents approximately 60% of the total primary income, with agriculture and fishing accounting for around 16 and 24% respectively.

In December 1995, LAET was approached by two leaders of the Porto de Moz District, who asked for assistance in organizing a seminar on the future of wood and fish in their district. Several popular organizations of this district, including the Farmers' Union, had already organized three meetings on these topics. During the third seminar, in which a MPST representative was present, participants decided that they would need some technical support from LAET to organize something bigger at the next seminar, which was planned for June 1996.

Based on their previous experience in Uruará, the researchers, together with the MPST, adopted a different strategy for participatory planning. Rather than inviting all the local stakeholders and local technicians at the beginning, priority would be given in the first stage to a direct dialogue with the local rural people's organizations, since these were the people who most needed empowerment. Once their objectives and strategy were clarified, they could later enter into negotiation with other stakeholders and with the government.

The Rapid Natural Resource Appraisal

During a first visit in March 1996, the LAET coordinator, together with an MPST representative, discussed how this support could be organized with local leaders. Representatives of communities in the district were planning to be in the district headquarters during one week in May for a religious meeting. To take advantage of their presence, it was decided that LAET would ask them to help conduct a "rapid natural resource appraisal."

The participatory research was relatively well prepared and the whole LAET team was in the field for 10 days. They quickly established a picture of the district's social, economic, and ecological dynamic. More importantly, these results where quickly analyzed and preliminary results were presented to the committee members one week before the seminar. It was then possible to have a preliminary discussion of the results and answer such questions as: "Did the information correspond to everyone's personal impressions?" "Was the information clearly presented?" "Was this the best moment to discuss certain problems?" and "Would it be better not to mention them at that stage?"

This preparatory meeting also was an opportunity to discuss the planning and facilitation of the seminar. The local leaders clearly wanted to be in command, and they wanted this to be known to all participants. But this common planning helped them to better organize the schedule, to discuss who should present what at which moment, etc., and to leave sufficient time at the end for discussing concrete actions. The MPST representative was particularly efficient in this task. He constantly served as a link, sometimes practically as an interpreter, between the local leaders and the research team.

The results of the participatory research were presented at the municipal seminar in June, with participation of more than 80 representatives of the district communities, some representatives of other districts, local technicians, and a representative of the State Secretary for the Environment. It also included the occasional participation of the district mayor and of a state deputy.

Major themes were (1) that large quantities of trees were harvested and exported outside of the district and (2) that returns to the district were low compared to returns that could be obtained for sawed lumber. It was estimated that the most accessible forests (e.g., less than 10 km from a river) would be exhausted in 10 to 15 years if nothing was done to curb indiscriminant logging. It was probable that the local dwellers would not easily find an alternative source of income (Rocha et al. 1996). The community representatives made various statements about the existing links between forest exploitation and reduced abundance of fish. Fishing was mostly important in daily nutrition, but was becoming increasingly difficult due to severe competition from commercial fishing boats, most of which were coming from distant places such as Abaetetuba (near Belém).

Another theme was land: availability was becoming critical. Most traditional occupants never thought of registering land until 10 years earlier (about 1986). Land (and its natural resources) was practically in free access. Recently, however, both big and small loggers had begun to mark the boundaries and occupy vast areas of land that they claimed as theirs. Relatively small areas were left for the local inhabitants, who were instructed by the union and the Catholic Church to register a standard area of 50 hectares per family for agricultural colonization. As a result, various communities were circled by logging companies and pressured to sell their remaining lands. A quiet but intense debate then occurred in small groups of community representatives. Special care was taken to avoid external interference as much as possible in these small groups, and technicians and researchers were conveniently confined to a special group of their own. The community groups listed proposals, some of which were quite ambitious (e.g., the founding of an environmental commission in the Municipal Council, reinforcing control on the taxes collected on log exports, creation of a district tax on logs exports, and so on), but others were quite concrete, such as the establishment of community fishing rules and their immediate enforcement.

The Natural Resources Management Program

After the seminar, a municipal committee was formed to develop a natural resources management program. This program included an agreement between MPST and LAET to work together to select priority areas and to implement recommendations of the seminar during the next three years. Some important results of the program were:

- The rapid spread of community rules restricting fishing in their rivers and effectively gaining control of the commercial fishing in their areas
- Start of the discussion of community forest reserves in four communities and demarcation of boundaries in two communities
- Organization of a program of environmental awareness to explain conclusions of the seminar (a small illustrated booklet and posters were produced) and a discussion of existing laws and authorities that could be approached
- Creation of a protected area in a varzea (a seasonal lake called Urubu), a plan officially backed by the State Department of the Environment, despite serious resistance from a large landowner who claimed that this lake was his property
- Support obtained from the Federal Environment Agency (IBAMA), which made at least two visits to the river and confiscated all fish from an illegal commercial boat
- Partial support obtained from the Public Land Office (ITERPA) to grant access to its records to the Farmers' Union and to change its local representative, who was considered corrupt

Discussion

In Porto de Moz, the platform approach for natural resource management was applied differently from what was done in Uruará. In the first stage at Porto de Moz, not all stakeholders in the PAR exercise were involved. Priority was given to the rural poor and their organizations. Only when these groups and organizations were reinforced and had a clear idea of their own interests and strategies was a negotiating platform tried. As a result, more concrete results of natural resource management were obtained in two years at Porto de Moz than in four years at Uruará. It can be argued that these results can be attributed to the different cultural context of Porto de Moz. The local communities are more structured in Porto de Moz than in Uruará and the peoples' link with their land is stronger, since they usually live for several generations in one place (or at least in the same environment).

In Porto de Moz, however, the rural peoples' organizations were weaker than those in Uruará. In Porto de Moz, traditional relationships between merchants and their clients, based on permanent debt, are still very strong. The fact that concrete results were obtained despite this paternalistic relationship is certainly a significant result.

An important aspect of the experience is that some of the results were obtained through direct contacts of the local organizations with the national administrations. This contact was greatly facilitated by MPST support and experience in this area, but also by the fact that the participatory research results were quickly published and communicated to decision-makers at the national level. This confirms Sawyer's (1990) theses that states that results in natural resource management in Amazonia can be better obtained through direct cooperation of the local organizations with national administrations, thus bypassing the local elite, who are likely to block all initiatives that will go against their immediate economic interests. If this is true, it means that platform negotiation involving all local stakeholders is not presently the best approach for natural resource management in Amazonia.

Another important result of the Porto de Moz experience is that the farmers' organizations played exactly the roles that were expected of them; that is, they facilitated research, helped to understand the farmers' strategies, pressurized the research toward client-oriented priorities, publicized the results, and represented the farmers in negotiations with the state. It is difficult to imagine a meaningful participatory planning effort at the municipal level without the full involvement of the local farmers' organizations.

Finally, one can observe an ongoing learning process between the researchers and the farmers' organizations. The division of responsibilities and roles among LAET, MPST, and the local leaders was much better defined in Porto de Moz than in Uruará. A common strategy between MPST and LAET was clearly spelled out from the beginning and integrated some lessons from the previous Uruará experience. A specific written agreement, discussed after one year of experience, helped to formalize a medium-term strategy with the local organizations and reduce the risks of drifting away from the agreed-upon agenda.

Case III: Altamira

The Farmers' Union Research

In 1996, the Farmers' Union of Altamira District approached LAET to obtain support for an agricultural survey they were planning to conduct in Altamira and three neighboring districts. They wanted to gather their own data and statistics about trends in the rural area so they could have a more solid foundation in discussions with the local councils and the national government. The generally available public statistics are notoriously inefficient or biased in Amazonia. The directors of the Farmers' Union made it clear that they wanted to be in charge of this research. The role of LAET was restricted to technical and methodological support in terms of preparing a questionnaire, training the young farmers who would conduct the survey, and helping in the data treatment. In return, LAET was granted free access to the data that would be obtained. The proposed effort was quite ambitious—a general survey of all families living in the countryside and a detailed survey of one-tenth of these families.

The detailed questionnaire covered most aspects of the lives of the farmers—from their previous history, land-use pattern, agricultural production, and income estimate to social life and children's education and future. Forty farmers (mostly young, educated farmers' sons and daughters) were trained to interpret and apply the questionnaire. They then went on to conduct the survey, traveling by foot or bicycle, in the areas that they knew best. The result was surprisingly good in actual coverage of the area and families and in the quality of data obtained. The gains in the confidence of the farmers and the truthfulness of their answers to the well-known local inquirers largely offset their lack of technical skills. The results were quickly computerized and analyzed with the participation of the most capable local investigators, then were presented to the communities. A systematic treatment of this information, combined with the use of recent satellite pictures to detect the new areas of agricultural expansion and deforestation, is underway. Results should provide a sound quantitative database to test many of the hypotheses that were formulated in LAET's regional diagnosis.

Land-Use Planning

The Altamira Farmers' Union was also involved with a group of local organizations interested in the use of the Altamira District land in the south of the município. The union proposed a general land-use planning of the district, using the participatory methods developed by LAET. This particular activity, still in progress, is interesting for the following reasons:

- The demand was formulated by the farmers' organization at a time when criticisms against the inefficient research of public agencies and their lack of support for the farmers' organizations were highest. It confirms that the information produced by research is important to the farmers' organizations in their negotiations with the public sector, and that they intended to maintain control and benefit as much as possible from this new asset. Furthermore, they want to be trained to increase their capacity to conduct other such research when needed. This is clearly a process of empowerment for the local organizations, a process that the organization itself initiated. In spite of all the rhetoric of local participation used by LAET, their researchers had never thought that such an initiative could occur.
- It was easy to outline an agreement on a general land-use planning activity. A written agreement, designed in the early stages, discussed the responsibilities of each party and the subsequent ownership and use of the collected data. The farmers' organization insisted on presenting the results to their membership and to local technicians and government officers. LAET's interest was in further analysis and publications.
- This type of farmer-led research can help to solve the contradiction between the pressure on researchers to provide more immediate results and their need to collect quantitative data suitable for publication.
- The participation of the Farmers' Union in the municipal-planning proposal was encouraged by the fact that the farmers could already demonstrate their own competence and mastery of the participatory research method. Their confidence in the data collected and their willingness to discuss it with researchers have also increased.
- The idea of popular participatory planning and land-use planning is gaining ground in the region; it is not impossible that in the relatively near future farmers' organizations will take the lead in proposing

participatory land-use planning and participatory research on land use in their districts.

Lessons from the Case Studies

The Uruará experience was disappointing in its lack of concrete results, but was rich in observations and discoveries.

In a context in which, for practical purposes, the law and the state are absent (or interpreted and used by the most powerful in their own interests), it is difficult to negotiate and carry out a plan based on the notion of "the interest of the greatest number" and to guarantee that decisions taken collectively shall be effectively implemented.

It is important to distinguish between the rhetoric and the real objectives and strategies of the actors involved, especially in a cultural context in which dissimulating real objectives and fooling other participants are seen as a sport and a motive of pride. In Uruará, it has become clear recently (at least to the researchers) that the main objective of all participants in practically any kind of project was to attract funding from the national government and that few people actually believed in their own capacity to solve at least some problems locally without external funding. Another hidden agenda in the planning exercise was to prepare a new group of candidates for the next local election. These candidates were supported by an alliance between some representatives of the farmers' groups and some local businessmen and sawmill owners.

Local technicians working in the state agencies (especially in extension services and regional development agencies) and outside researchers invited for their expertise also had their own personal agendas, sometimes with an interest and strategy in local politics. Often, they were not really completely objective representatives of the government. Rather, they are actually also "stakeholders." This causes an additional difficulty in the negotiating process, because they are both judges and parties in the negotiation process and tend to use their position to defend personal interests.

Recommendations

As a result of the experiences with the multiple stakeholders platform method at Uruará, Porto de Moz, and Altamira, we can make the following recommendations.

1. Any person or entity who wishes to assist in a planning process such as that in the Brazilian frontier should be able, at a minimum, to

analyze and understand each party's hidden strategy and interests. This means having on the team a person—preferably an anthropologist or sociologist—with extensive field experience, especially in analysis of local politics. In addition, at least one team member should reside in the community, even if he or she has no social science background.
2. In the frontier, it is not possible to use Western-based concepts of conflict resolution through negotiation and discussion of all concerned parties. One cannot expect a final decision by the courts or enforcement by the government to guarantee the completion of the consensus agreement, at least not within a short-term or medium-term perspective.

Once the latter realities were understood by the research team, an option was to forget the idea of local planning and to concentrate on the *reinforcement of democracy* and the *establishment of reliable law enforcement* within the local society. Obviously, this means adopting a minimum of 20 years' perspective (the next generation) and hoping that there will still be some forest and small agriculture to be saved by then.

The other option was to experiment with *new methods and paradigms of local intervention* for participatory planning. LAET reached the conclusion that a possible strategy for such an intervention is to first establish an alliance with one of the parties (which means spending time and effort to build confidence and mutual trust) and then to discuss a strategy and assist this party in a negotiation with other parties (but not necessarily all other stakeholders) or in the formulation of an independent project of natural-resource management. One might discuss which is the best party to choose. It might be easier to form such an alliance with powerful groups (such as sawmill owners), who have the capital and political links that facilitate discussion of new forms of management. However, all indications are that a large number of the local elite has no interest in land-use planning, because they derive part of their profits from the free access situation: free wood extraction, cheap pasture establishment, or land speculation.

The most productive approach from the viewpoint of natural resource conservation seems to be to reinforce the farmers' organizations, so that they can use their political strength (in great part derived from the electoral weight of the rural majority) to outweigh the interests of the local elite. The chance that farmers will effectively support a land-use policy restricting access to new areas depends on government willingness to give additional support in infrastructure and education to regions with populations of small farmers. Participatory land-use planning also opens the possibility of the creation of community or municipal forests, as initiated in Porto de Moz.

Conclusions on the Platform Method of Participatory Planning

The following conclusions refer to the initial propositions on which the method is based:

1. It was confirmed that the local stakeholders were able to make various proposals that would be in the majority interest and would reduce the overexploitation of natural resources (e.g., the establishment of a municipal forest, the protection of some fish reproduction areas, and the taxation of timber exported in logs).
2. Participatory research was found to be an efficient method of integrating new ideas and elements of diagnosis into the public debate. It was accepted because it was clearly the result of "summing up" the popular knowledge with a scientific approach. As a result, it changed the power relations in the district and empowered those groups who, although in the majority, nevertheless were dominated and exploited by the elite power structure. However, facilitation was best exercised by the farmers' representatives themselves, when a straightforward working relationship with the researchers was guaranteed.
3. The State did not play its role in guaranteeing the effective application of measures proposed by the majority (and even approved officially by unanimity in Uruará). In fact, elite groups were able to manipulate the process of multiparty negotiations, so the majority had little influence on the final outcome of the process. The multiple stakeholders platform method and conflict-management approach have been developed mostly in the context of developed countries and are based implicitly on a reliable system of justice, law enforcement, and widely recognized democratic concepts and values. LAET found that these approaches were inefficient in the frontier context marked by paternalistic values, limited authority of the State, and its appropriation by the elite (patrimonialism). This conclusion could probably be extended to the situation of many developing countries with similar characteristics.

An ethical problem must be mentioned at this point: there is a risk that the local elite, threatened in its interest, will respond by using violence. The history of the Seringueiros Movement and of the assassination of Chico Mendes shows that this is a distinct possibility in Amazonia. Researchers must be cautious in this situation, not so much for themselves, but for local leaders. Radical proposals, which would greatly affect the elite interest in a short time, should be avoided.

The Potential of Participatory Action Research for Testing Methods

An important aspect of PAR is to use feedback from the initial research-action cycle to identify mistakes and correct actions to gain improved results during a second cycle. This is exactly what happened during the succession of municipal planning experiences described here. The Uruará case tested a method of local planning, which had been widely discussed by Amazon specialists but never put into practice. As a result of this case, limits and weaknesses of the method were analyzed, and the wisdom gained was incorporated into the second case at Porto de Moz and then in Altamira. This is a clear demonstration of the potential of PAR as a scientific and pragmatic approach to solving problems of natural-resource management.

PHOTO ESSAY

Deforestation of the Brazilian Rainforest

1. Aerial view of the mature forest near Altamira, Pará, Brazil.

2. Track (picada) through the forest made by loggers to extract timber.

3. Logging truck near sawmill.

4. Colonist's house in an agrovila, a small village, remotely located.

5. Rice field after harvest by colonist farmer; the field was originally forest.

6. Farmer's family threshing rice.

7. Farmer with diversified crops and holding a root of cassava (*Manihot esculenta*), another common staple.

8. Oxcart used to haul crops to the village.

9. Transamazon Highway in repair near Altamira.

10. Forest recently cut and burned for conversion to pasture.

11. During the first three or four years, pasture production is good. A Zebu variety of cow from a large ranch is shown here.

12. After several years, woody shrubs invade the pasture. Fire is used to control the shrubs.

To Decrease the Pressure to Clear New Forests, a PAR Program was Adopted to Encourage Farmers to Regulate the Use of Common Resources and to More Sustainably Manage Their Land Already under Cultivation

13. Headquarters of the MPST (regional farmers' organization).

14. Meeting of MPST members to discuss credit programs.

15. Research team from LAET carrying out a survey of farms in the region.

16. Wood-sawing trial (in Uruará) to show farmers the importance of value added at the local level.

17. LAET trainee explaining results of a zoning survey.

18. Rubber tapper. Reserves for such extractivist activity were proposed as part of the zoning plan.

19. Zoning is difficult to enforce in frontier areas, where ranchers and loggers oppose any land use regulations.

20. In established riverine communities, such as Porto de Moz along the Xingú river, there is a greater interest in conserving natural resources.

21. Meeting of the local communities organized by the muncipal Natural Resource Committee in Porto de Moz.

22. House of fisherman near Porto de Moz, where fishing regulations have been established and enforced.

23. Use of animal traction is one of the ideas proposed by LAET for the small farmer.

24. Bee-keeping is another potentially profitable enterprise for small-scale farmers, but marketing the honey is a problem.

25. The availability of primary schools such as this one (also used for community meetings) is important to farm families.

26. House of small farmer in remote area with intensified production of a variety of crops.

27. Agroforestry system, where species with varied ecological niches are mixed together to promote more efficient use of light, nutrients, and moisture.

28. Pueraria planted as a cover crop to restore fertility through addition of carbon and nitrogen, and to eliminate weeds.

29. Legume interplanted with coffee to enrich the soil and reduce weeding.

30. Coffee seedlings in nursery organized by the farmers' association.

31. Cacao plantation, showing fruit.

32. Mahogany tree planted within cacao plantation.

33. A 25-year-old mahogany within a cacao plantation.

34. Plantation with Brazil nut overstory and cacao understory (EMBRAPA research station, km 23).

35. Plantation with rubber overstory and cacao understory (EMBRAPA research station).

36. Future generations along the Xingú river.

Results at the Farm Level

Joint LAET-MPST (Laboratorio Agro Ecologico da Transmazônica-Movimento Pela Sobrevivencia da Transamazônica) activities occurred at two levels. The *communal level* was discussed in the previous chapter. The *farm level*, introduced briefly in Chapter 5, was targeted at the individual/family level of resource management. The timeline for these farm level activities was presented in Table 5.1. In this chapter, we present results in two of these farm level activities: (1) research development on perennial crops and agroforestry and (2) the discussion of a more appropriate credit policy for individual farmers.

Research Development on Perennial Crops and Agroforestry

The first agricultural survey and initial discussions with the farmers' organizations confirmed the importance of establishing perennial crops (mostly but not exclusively tree crops) for the stabilization of the colonist agriculture. For the three main perennial crops of the region (cacao, black pepper, and coffee), the average income per unit of land is higher than for any other agricultural use. The return to labor is also better than that for annual crops and roughly the same as that for beef cattle. A farmer who invested in perennial crops does not need to clear new areas for rice or pastures. He can concentrate most of his workforce on the crops and expect a reasonable income.

The survey found that the most successful farmers were those who diversified their production. They divided their labor between perennial crops and cattle, and they grew annual crops only for family consumption. In addition, farmers who had perennial crops were likely to make efforts to avoid uncontrolled fires by limiting their own pasture area, leaving natural vegetation as firebreak, maintaining artificial firebreaks, and controlling the neighbors' practices.

Another big advantage of many perennial crops is that access to the market is not limited. These commodities can be stocked for a period of time and have relatively high value per kilogram, so that the cost and timeliness of transport are not limiting, as with fresh or bulky products. It is always possible to sell these products to middlemen or big merchants. On the negative side, the interannual price fluctuations are very high, owing to world market fluctuations. Price varies by a factor of 5 between peak and low years. Worse, these commodities follow a 5- to 10-year market cycle, so farmers have to bear low prices for several consecutive years. During periods of low prices, farmers tend to abandon their permanent plots and invest more in cattle and annual crops. Although some farmers considered organizing themselves in marketing cooperatives, a market study showed that not much is to be gained from this, because world market prices were the main factor behind the fluctuations and the margin of intermediaries was not as high as farmers believed.

Another serious problem affecting plantations was the occurrence and spread of various diseases, particularly the *Fusarium* wilt in black pepper and "witchbroom" fungal attacks on cacao. As a result of these diseases and other management problems, farmers' yields of cacao, for example, varied from 200 to 1200 kg per hectare (potential yield in the region is higher than 2000 kg/year). In an effort to establish a more stable yield of perennial crops, farmers and technicians tried various novel types of interplantings, such as an association of cupuaçu and coconut. However, the problem here was that the market for many of these products was far from guaranteed. There were many instances in which fruit production could not be sold because the cost of transportation was higher than the market price.

LAET also conducted a survey of farmers' innovative practices in agroforestry, which showed that most existing combinations of interplanted species were organized around one of existing commercial crops of fruit or timber trees. Therefore, LAET decided that it was more urgent to concentrate research on the existing tree crops than to engage in new agroforestry research.

Two specific agronomic studies were conducted on black pepper production in 1995. The studies indicated that capital and technical knowledge were the most important factors in explaining the great differences in yield and economic return between different farmers. Soil fertility had less of an influence. In the case of black pepper, a critical point was to avoid transmission of the disease by the apparently healthy plants obtained from cuttings. Most farmers were unaware of this. When they learned they were unwillingly accelerating the spread of the disease, they requested LAET to help them solve the problem. It took some time before LAET succeeded in mobilizing an EMBRAPA (Empresa Brasileira de Pesquisa Agropecuaria [National Agricultural Research]) specialist to assist in a program of technology transfer and training with three groups of five farmers, selected by three farmers' unions.

Farmers in the program were trained to produce healthy pepper plants that they could first plant themselves and then sell to other interested farmers. The results were so impressive that it created a big demand from other farmers. This activity started in 1997 and, by 2000, some farmers had begun to achieve economic results.

The potential is very high in the region. If farmers are convinced that they can control the disease, more than half would grow black pepper rather than invest in pastures. In addition to working on disease control, some farmers, accompanied by LAET, started to experiment with leguminous cover crops between the rows of black pepper with good success. Instead of hoeing, weed control was ensured by the legumes themselves (especially with feijão de porco, *Carnavalia ensiformis*), which were regularly pruned to facilitate passage between the rows. Soil fertility was notably improved. Some farmers claimed that they got better black pepper yield, but this still needs to be verified by adequate on-farm research. This innovation can reduce the disadvantages associated with black pepper cultivation and make it more sustainable.

Conclusions about Research and Development of Perennial Crops

Agroforestry is generally considered an appropriate alternative for Amazonian agriculture. Structures and functions of agroforestry systems resemble those of the native forest. These structures and functions help conserve nutrients through efficient recycling, a factor that is important in Amazonian ecosystems where the potential for leaching, volatilization, and erosion of nutrients is high. Various research programs have been conducted to test the technical and ecological sustainability of agroforestry systems, including work at EMBRAPA stations. However, experience has shown the need to establish the economic viability of agroforestry systems before doing research on their technical aspects. After the International

Institute for Tropical Agriculture (IITA) made huge efforts to design new agricultural systems for the humid tropics, the same conclusion has been reached regarding the use of alley cropping (Lal 1991). This illustrates the fact that most monodisciplinary and top-down research efforts are unsuccessful in introducing new technologies in the farms, including agroecological innovations, a point that has led to alternative development initiatives such as participatory action research (PAR).

PAR showed that significant improvements in sustainable land use can be achieved through developing existing permanent commercial crops in the region, even if they are not usually considered as agroforestry. This is because their structure and function (except for black pepper) resembles those of agroforestry systems. Black pepper does not ensure full vegetation cover and is frequently cultivated in a manner that exhausts soil fertility, with three or four hoeings per year to keep the alleys between the rows "clean" from competing weeds. All perennial crops promise good economic returns to farmers and employ large amounts of agricultural labor, therefore reducing the pressure to establish pastures. Participatory research invested in these crops is likely to be much more efficient for developing sustainable agriculture in the region than for research invested in new agroforestry combinations, which satisfy the minds of researchers but are of limited economic interest to farmers.

This example also confirms the importance of farmers' organizations, to pressure researchers into recognizing farmers' priorities. Regular reality checks by researchers is an important element in building a dialogue within PAR. The examples also show that the action research team cannot function independently from the rest of the technoscientific world. Once a specific problem is clearly stated and its solution is in view, the action research team needs to obtain the cooperation of technical specialists.

The Credit Debate

Credit is a powerful tool that small farmers can use to stimulate changes in agriculture. It can either encourage large-scale forest conversion to pasture and favor further land concentration or help small family farmers survive on the basis of a more diversified agriculture. Given the number of farmers benefiting from credit in the region, any improvement in credit rules regarding sustainability would have a sizable impact at the regional level.

One of MPST's main accomplishments in 1991 and 1992 had been the opening of a special line of reduced-rate credit for small farmers. Land title was not required. In 1993, this credit was extended to dairy cattle development and an agroforestry plantation composed of cupuaçu and coconut. The farmers had to choose among various technical packages ("tech-packs"). Choices were to establish either (a) 4 hectares of the agroforestry system; (b) 1 hectare of agroforestry and 10 heifers; or (c) 10 heifers, a bull, and some wire to fence pasture. To obtain the credit, the farmers had to join a small farmers' association and obtain a certificate from the Farmers' Union, stating that they were legitimate family producers. Specific limits for the annual income, area, and number of paid workers were also set. Apparently, the suggestion that the credit should encourage agroforestry came from both bank officials and farmers' organizations. To avoid criticism from the public opinion and environmentalist organizations, both the bankers and the farmers wanted to show that the credit would help stabilize the family farmers' agriculture. Since both the banks and the farmers' organizations had only vague ideas about the nature of agroforestry, they turned to the official research—EMBRAPA—for suggestions on the various possible agroforestry designs.

EMBRAPA supplied several combinations of species with their recommended spacing, such as seringa (rubber trees) and cacao or coffee and black pepper, and so on. From this shopping list, the farmers' representatives and local technicians chose the suggestion that seemed to be best adapted to the region. A frequently chosen combination was cupuaçu and coconut, because it was anticipated that the market for cupuaçu would grow quickly. There was also pressure from the bank, who had financed a local agro-industry for frozen pulp of cupuaçu and wanted to guarantee its future supply. Nobody apparently foresaw the problem with cupuaçu. It is a very soft fruit and has to be picked and transported to the factory twice a week. This means that an extensive transportation system was needed, since the farmers who would benefit from the credit would be scattered along thousands of kilometers of rough and sometimes inaccessible tracks. Also not considered was the potential use or market for coconuts in the region. The local market is limited. Since coconuts are bulky and cannot be economically transported very far, the only use was for farm production of poultry or pigs.

As a result of the availability of easy credit, there was a quick bloom of new associations, and many farmers benefited from this credit. As problems with cupuaçu became apparent, many farmers chose to develop their beef cattle herd. As a result, there were some important effects at the regional level. The abandonment or sale (both of which were common before 1991) of land was halted. There was a resurgence of land prices and a stimulation of the beef market, especially for calves and heifers. MPST benefited from these trends through increased recognition among farmers.

In 1994, a survey of 55 farms in the region by LAET researchers showed an alarming trend toward massive conversion of forest to pasture, even in lots where no cattle were yet present (Castellanet et al. 1995). These results were presented in a small, illustrated, easily read booklet, which discusses the future of the region's agriculture. During a discussion of these results at the PAET (Programa Agro-Ecologico da Transamazônica) annual meeting in November 1994, the risks associated with this trend toward pecuarização (development of beef cattle production) were reported. Some of the risks were environmental (risk of landscape degradation) and some were social (risks of land concentration and substitution of family farms by big ranches). Many of the farmers' representatives argued that cattle constituted the only economic option in light of the crisis of other commodities, especially pepper and cacao. Nevertheless, other farmers supported the idea that a more diversified and intensified agriculture should be encouraged.

In August 1994, the MPST administration asked for specific research on the impact of credit on farmers, so that they could assist them in monitoring the results. This was part of a statewide effort of the Farmers' Union federation, to bring together information on the impact of credit on small farmers. LAET participated in various discussions on credit at the regional level but was slow in responding, partly because it had a lack of manpower and partly because it felt that such credit was a politically complicated matter. By the end of 1995, many associations were clearly not properly monitoring the application of the loans. The majority of farmers did not have confidence in the future market of cupuaçu and consequently did not properly care for their agroforestry plots. Also, some local leaders appeared to have found ways of obtaining loans for which they were not entitled. Some even stole a part of their association's loan money. Corruption was increasing and affecting many associations.

In many cases, the directors purchased poor-quality heifers on behalf of the members of the association. Kickbacks from the big ranchers were suspected. In the beginning of 1996, the MPST administration decided to intervene in the most severe cases and denounce the corrupt local leaders, but without external publicity.

Beginning in 1994, some LAET members participated in a "regional credit committee" together with the MPST, the extension service, and EMBRAPA. In June 1996, the MPST Board of Directors decided to organize a regional meeting on credit and asked LAET to present its findings to fuel the discussion. The team was cautious, not willing to be sacrificed as a scapegoat in this difficult question. After some delays, MPST's directors and LAET reached a mutual decision that LAET should first take some time to analyze the problem and let MPST decide the direction to be taken before calling a bigger meeting. A strategic credit seminar was, therefore, organized in October 1996 between MPST directors and LAET researchers. It helped to clarify the situation and their options.

Results of the Credit Seminar

Based partly on the research presented and partly on the farmers' own observations, the group, consisting of MPST directors and LAET researchers, came to the following consensus:

- There was a real economic risk for the farmer who produced only beef cattle.
- Many farmers did not have the experience and minimal technical skill to obtain good performance and financial returns from cattle rearing.
- The risk of the farmers' inability to repay their loan was increasing at an alarming rate.
- In contrast to earlier years, there were no recent signs that land holdings were being concentrated in the hands of a few speculators. However, the risks of this would certainly increase with insolvency and the end of the easy credit era.
- The agroforestry systems were generally not cared for properly or completely neglected.
- In the cases in which the cupuaçu was well established, the marketing was not guaranteed.

On the *organizational* and *political* side, the same group came to the following conclusions:

- Many local associations that were established to certify credit were either weak or existed only on paper.
- In many cases, their own leaders induced the farmers into credit-taking and noncontractual uses of credit.
- There was no individual guarantee of a loan. Many farmers were accepted in the association for the sake of friendship, whereas everyone knew that they were unable to manage their farm and reimburse credit in the future. Although local organizations and prefeitos were responsible, in part, for the failure of the Fundo Constitucional do Norte (FNO) credit program, MPST was first to receive the blame.

Gradually, the group came to agree on the following directions for the *future*:

- The credit should be more diversified and adapted to each farmer's needs and capacity. This meant special training for farmers' leaders and extension agents, and also a much more selective program.
- The responsibilities and duties of each local association should be spelled out, and MPST should refuse to work with associations that did not reach certain ethical and organizational standards.

The risk associated with this strategy was that the majority of farmers might decide to join other associations (set up by local politicians) with less rigid norms. But MPST leaders felt that it was not a very big political risk, since associations were weak anyway. Moreover, it was a lesser risk to MPST, compared with the criticisms that could arise from continuing the same paternalistic strategy.

Once this level of discussion was achieved, MPST leaders started to discuss possible consequences and reactions to this new strategy, and finally decided to organize a regional meeting in December 1996. Once again, the problems were presented and discussed alternatively by researchers from LAET or by MPST leaders. The group came to roughly the same conclusions and orientations as did the seminar, although some important leaders, including the State Representative, who was also an ex-MPST leader, clearly did not agree with this new orientation and preferred to continue the old "mass mobilization"scheme (mass mobilization referring to the strategy of accepting all associations, regardless of reputation). Nevertheless, the facts presented were sufficiently clear and convincing, and it was decided that the new proposal should be submitted to the annual general meeting of MPST in March 1997. However, after internal discussions, the subject was buried and never discussed again with LAET. Although LAET never got a clear explanation of what happened, they suspected that the recommendations were forwarded through the Farmers' Union hierarchy where, at the national level, they were considered "politically unacceptable," either because they implied a criticism of the leadership or because they had possible consequences on national credit policy.

Lessons from the Credit Debate

Several lessons can be drawn from the credit debate:

1. Although the credit problem was identified in 1994, it was useless for LAET to force a debate before the time was ripe and before MPST leaders were convinced that something should be done. This moment came as a result of the accumulation of small problems and also because of a change in MPST attitude. This change was partly in response to the attitudes and policy of the bank, the public opinion and newspapers that influenced political strategies, the creation of new organizations in the region, the outcome of local elections, the relations of MPST with the state Farmers' Union's new board of directors, and so on. It is difficult for the external researcher, even for a sociologist, to be able to understand the strategy of organizations such as MPST. If researchers are not accepted and trusted by organization leaders, they must be patient and

try to see the situation through the leaders' eyes. Researchers should learn to be patient even if they don't understand why things are not discussed as and when they would like them to be.

2. When the problem is considered ripe, then research input can be effective and important. This means that the team should continue to work on issues that are not yet a priority in the farmers' agenda, but might become so in the future. The question is not whether to do the research, but when and how to present and discuss the findings of this research.
3. Effective progress in such a delicate field such as credit requires a high level of confidence and trust in the research organization by the farmers' organization, so that realistic alternatives can be worked out together. In the case of credit, for example, it was probably unrealistic to consider that a major policy change could be assumed by MPST at the regional level. The stakes and risks were probably too high for the organization. In this context, an experimental approach, restricted to a specific group of interested farmers, would have been more appropriate to test what would be the impact and problems of a new credit policy for the farmers. This test could then be discussed at the regional level based on concrete facts. LAET and MPST did not even consider such an approach until 1997.

Evaluation of PAET from the Farmers' Point of View

Because MPST was concerned about the slowness in obtaining concrete results from PAET, they held an internal evaluation of the program in December 1996. This evaluation confirmed that direct results at the farmers' level were limited: the animal traction program was not giving satisfactory results, a farm management and accounting effort was considered promising but not sufficiently advanced to serve as a tool for advice on changes in the farmers' management, the wood-processing project had not gone forward, and the black pepper replanting program and the effort to make available leguminous plants for cover crops had just started. PAET certainly had an impact on the opinions of the farmers' leaders, by reinforcing their interest and concern about diversification and about the need to better adapt credit to the farmers' individual circumstances. However, the impact on lay farmers was limited. Relatively few farmers in the region knew about LAET, and an even smaller number had any contact with the research team.

On the positive side, the discussion of the value of wood products and the sawing test in Uruará had an impact on some communities, who decided not to sell their wood for low prices. It also encouraged various farmers to plant native trees (such as mogno and cedro) after they realized the value of the wood and the quick growth of these trees. In Porto de Moz, the municipal planning exercise had resulted in the defense of local fishing rights and also in the creation of community reserves and control of logging by local communities.

The Casa Familiar Rural school in Medicilandia was probably the greatest practical result of LAET from the farmers' point of view. Practically all other municípios of the region developed a great interest in this initiative. The opening of a new agricultural science degree (Licenciatura de Ciencias Agrarias), based on the Altamira campus in 1997 with a strong orientation toward family farmers in the Altamira region, was also seen by MPST as an important milestone for the future, but not as an answer to the farmers' immediate needs.

From MPST's point of view, PAET did not help to reinforce its organization significantly. On the contrary, some MPST leaders thought that PAET absorbed much of their time and effort but did not give much in return. As a matter of fact, LAET researchers did have a tendency to question MPST's projects and activities and often did little to improve them. In the elections, PAET support did not help to elect any prefeitos or members to the district council, contrary to MPST expectations.

On the other hand, the presence of researchers, together with the farmers' organizations in public events or in negotiations with other public agencies, certainly permitted MPST to gain prestige at a time when it was suffering from growing opposition from the prefeitos and big agriculturists.

The Learning Process

The impact of research action on personal change and capacity building is generally recognized (Lewin 1946; Liu 1992). However, this impact is difficult to quantify. Measuring direct results and impacts is only one aspect of the evaluation of a development program. If we consider, as PAR does, that development is essentially a learning process, one of progressively changing peoples' attitudes and aptitudes, then this learning process possibly needs to reach a certain stage in which a critical mass of people begin to think differently. Only then can measurable changes be observed. From this perspective, the fact that the farmers' unions started to adopt and apply themselves to the methods of participatory research and the fact that participatory land-use planning became a theme of public interest in the region can be considered significant advances.

The learning process also included LAET researchers. They often commented that they learned a lot from the PAR experience, especially about the importance of teamwork and having an interdisciplinary view. Researchers also learned much about the potential for, and the limitations of, change in the region. More deeply, they learned to maintain their professional standards while at the same time questioning their own work and scientific activities from the local people's point of view. It can be safely assumed that these new attitudes and capacities will be useful in other professional and geographical contexts, not necessarily restricted to the Amazonian frontier or research action programs.

Obviously, the critical test of sustainable development interventions, including training and capacity-building, is whether actual practices finally change as a result of these interventions. In the perspective of the learning process fostered by PAR, we expect that concrete results will multiply and become more significant in the near future for the following reasons:

1. An increase in direct work with lay farmers concerning appropriate technologies that have already been identified and tested, such as green manure for perennial crops; healthy pepper production; coffee production; timber tree planting; and improvement of cattle production through the use of minerals, basic sanitation, and pasture rotation.
2. A better and closer cooperation has been established between MPST and LAET in the areas of regional policies, especially credit; extension and training; land settlement and tenure; and creation of community reserves. This better cooperation can be based only on a better understanding of each other's priorities and strategy. Such understanding seems to be increasing. On the other hand, cooperation also depends on reestablishing mutual confidence and trust, which is not guaranteed.

Conclusion

The concrete results of PAR, in terms of changes in land-use practices, were still too limited to reach a conclusion on the efficiency of the approach in natural resource management. However, a reasonable optimism for future results, if the program continues, seems justified by the following observations made in 1998 and 1999 during various visits of one of the authors in Altamira:

- Promising technologies, such as the black pepper replanting method, have been identified and have started to be spread, with an important potential for improving the sustainability of small farmers' agriculture.
- Revolutionary concepts, such as the participatory land-use planning and community reserves, are being adopted and have begun to be implemented by the local organizations and some public services in the Altamira region.

In our view, these results justify the continuation of the PAR experience in Altamira and the starting of new programs using the PAR/NRM (participatory action research/natural resource management) approach in other contexts. In the same way that private industries invest a part of their profits in research and development, which gives little immediate return but may increase future production efficiency, national and international development and environmental agencies should also invest part of their resources in testing and developing new methods of intervention.

PART

III

LESSONS FROM THE PARTICIPATORY ACTION RESEARCH IN THE TRANSAMAZÔNICA

The Relationship Between Farmers and Researchers: Why There Was No Common Strategy

Although progress in many farm-level activities has been modest, some of the most recent initiatives have begun to justify the assumption that farmers' organizations can help in the participatory action research (PAR) program. However, the assumption that it was possible to define a common strategy between the farmers' organization and the research team was not verified. The fact that LAET (Laboratorio Agro Ecologico da Transmazônica) did not succeed in establishing a common strategy with MPST (Movimento Pela Sobrevivencia da Transamazônica) could be due either to the fact that MPST had no real interest in a better natural resource management in the region (a negation of assumption 1) or to a failure in the constructivist model of establishing a common strategy through improved communication. Here, wc cxaminc both possibilities.

Lack of MPST Interest in Sustainable Development and Better Management of Natural Resources

Given that the farmers' organization proved to be only marginally concerned with better management of natural resources (at least as understood by the researchers), it is useful to examine, with hindsight, the following questions:

- What is the interest of the farmers' organization in development in general? Do farmers have other objectives? Do these other objectives conflict or complement the development objective?
- Within the development objective, what priority is given to short-term results over long-term results? What is the relative importance of ecological or sustainable natural resource management concerns over other social or economic objectives?

Objectives and Functions of Farmers' Organizations: The MPST View

The MPST was created as a protest organization, whose function was to organize and mobilize farmers to negotiate with the government. Militants of the farmers' unions, who were also active members of leftist political parties, founded MPST. These members were mostly in the workers' party (Partido dos Trabalhadores), founded in 1980, whose candidate was almost elected president in 1990. However, the MPST did not consider that its mission was limited to political action. They felt that they had to gain credibility both from the government and from farmers. This entailed putting up "alternative economic projects," thereby demonstrating to the government the region's potential for development. It also entailed creating links with research to be able to carry out, for example, proposals to improve farming methods. MPST also felt that it should reinforce and increase the number of local farmers' organizations. This meant entering into groundwork, training, education, and facilitation.

The Call for Preservation by MPST: Rhetoric or Strategic Position?

Even during the initial stages of negotiation with MPST, there were doubts within LAET about the seriousness of MPST's interest in preserving natural resources, even though their motto since 1991 has been viver, produzir, preservar (to live, produce, and preserve). LAET later recognized that MPST had used this motto as a "flag" to improve their public image in Brazil. The fact is that, for most Brazilian citizens, the National Press, and particularly educated citizens with a progressive profile from southern Brazil, the colonization scheme in Amazonia is seen as a failure, an ill-conceived project of the military government intended to reduce internal opposition, and a useless waste of natural resources. Many people believed that there were no more small farmers in these areas and that the road was returning to the jungle (it actually happened in the eastern portion of Transamazônica, east of Itaituba). Therefore, some consensus existed throughout Brazil, after the return of democracy in 1984, that it was better to forget about these older colonization programs and to subsequently abandon all support for them.

The MPST's strategy was to first demonstrate that there were still many small farmers in these areas, suffering from the consequences of this ill-conceived program (another slogan at this time: "Transamazônica: it was a mistake to open it, it would be a crime to abandon it"), but who could practice environmentally friendly agriculture if given the chance. Second, MPST was to show that the large-scale ranchers, not the small farmers, were the main culprits responsible for deforestation. One of their actions was to organize a campaign for the survival of the Transamazônica, which culminated in a 2-week camp-out in front of the Congressional Building in Brasilia.

Even though MPST's environmental rhetoric did not seem genuine, keep in mind that no formal organization can function effectively by simply reflecting the sum of its members' individual opinions, especially when it has to negotiate with other organizations or with the government. The exercise of democracy is about explaining new positions and presenting "packages" of acceptable concessions in return for valuable gains that can be supported by the majority (Susskind and Cruikshank 1987). Therefore, it is an oversimplification to declare that the ecological positions of MPST are not genuine. Representatives of an organization must have leadership that is able to negotiate with the rest of the society, which entails having a broader world view than the rest of the membership.

As it turned out, MPST did conduct activities and projects that were really aimed at sustainable development, such as the launching of the Casa Familiar Rural (rural school) or the support of trials to use draft animals. It also played a role in the denunciation of illegal logging in the Altamira region, supported the recognition of indigenous land, and even proposed, in 1998, to create a regional forest on lands illegally occupied by large sawmills. At the same time, MPST was conducting activities more akin to mass mobilization and political conquest. In some cases, these two lines were in conflict, as in the Uruará municipal planning exercise discussed in Chapter 9, which eventually led to a falling out of MPST directors. One can consider that it is possible, and even desirable, to conduct basic development and political activities jointly, if such activities are based on a vision of development and democracy.

Farmers' organizations do not function as ideally as we could expect. Besides the fact that their objectives are not always clear and univocal, they are also composed of people who have their own personal interests, which they frequently give more importance to than the common interest. This can be true of organization members as well as leaders. In the Mexican farmers' organizations (Fox 1992), the decisions were rarely made during formal meetings, but were often the outgrowth of informal talks and debates of the leaders with people from their network of influence. Internal democracy and leaders' accountability increased or decreased depending on many factors, particularly the opportunities for rank-and-file members to meet outside the leaders' control. Such informal talks constituted a field (social space) in which one could promote a discussion about common interests and common future.

To promote this type of informal discussion, two conditions must be met: time and trust. Without trust, certain debates are censored before they can develop. Without time and patience, the researchers will try to impose their own timetable and force a discussion when it is not "mature." This simply means that a significant proportion of the public and leaders who are concerned about a given question are interested and open to discussion.

It is possible to understand, in hindsight, the tactics of the MPST leaders toward the members of their organization. They saw that if they planned some type of collective activity that would benefit the majority (considered a "gain" obtained from the government) and that activity had certain questionable aspects, this would indicate that the time would be right to start an informal debate on the questionable aspects.

The Farmers' Perspective

Sustainable Development

In many cases, a contradiction or conflict exists between the short-term individual interest of the majority of farmers and their long-term interest as a social group. However, this contradiction is not seen immediately, nor is it obvious. Some leaders, for example, never agreed that encouraging livestock could later result in concentration of land ownership. Others felt that forest exploitation was not a concern of the farmers' union. Even if MPST leaders have a long-term perspective on the future of the region, and they design proposals on natural resources management that are advantageous for the future of family agriculture, they will not get support from their constituents to implement them unless a majority has an interest in the long term. Even with a majority, much explanation and discussion are necessary, because members first have to agree on their diagnosis of the situation and then on a prognosis of the future.

However, there was evidence in the Transmazônica region that environmental concerns do exist within farmers' organizations, even if they have nothing to do with the researchers' concerns about biodiversity or global warming. The concerns were centered more around the future of farming in the region and their children's place in this future.

The long-term questions that emerged regarding the farmers' organization were:

- What proportion of its members are planning a future for their children in the region and are willing to cooperate with others for this future?
- What time frame is adopted by the organization for planning its strategy?

The Role of the Researchers

The farmers' organization initially had a twofold vision of the researchers. One view was that ordinary researchers (e.g., agronomists) could help to introduce new technologies (e.g., small generators) into farms to increase agricultural production or improve daily life. The other view was that ideologically committed researchers (e.g., ecologists) could help farmers to gather information to support their claims and prepare projects for government or funding agencies. Such researchers can also help them as advisers in analyzing specific problems. Apparently, the farmers did not consider this type of advice to be part of "development" work, but rather a "political" activity.

The Meaning of Development

As a result of the evaluations performed by LAET, it gradually became clear that the farmers' vision of development was different from LAET's view. For the farmers, development primarily meant improving living conditions for their family through increased agricultural production and social benefits resulting from government subsidies. For various farmers' leaders, the only important parameter was the result, independent of the way these improvements were obtained. On the other hand, LAET had a broad definition of development, which considered education and training not only in technical aspects, but also in ecological terms: sustainable ecosystems are the foundation of sustainable development. As the program developed, LAET's concept of development became even broader. Initially, the researchers were concerned about compatibility between the political function of farmers' organizations and their technical function. They wanted to make a clear distinction between both, because they could not participate in political activities but could cooperate in more technical matters. Later, the researchers came to see "political education" as a premise for any long-term effort for better management of natural resources. However, for LAET, "political" did not mean entering partisan politics and campaigning during elections, but rather *increased participation of citizens* in the public debate about their common future and about the rules and means to reach a commonly defined goal. It meant *promoting active citizenship and democracy*. Political education also entailed supporting better functioning of the existing organizations, including training of present and future leaders of farmers' organizations.

Another area of misunderstanding was in the model of development to be sought. Local organizations had a tendency to expect the State to transfer more resources to local farmers. However, farmers never questioned where the additional resources should come from. Furthermore, they were under the impression that funds obtained from external nongovernment organizations (NGOs) were supposed to be renewable and could be used in any manner, regardless of the project for which they were originally designated.

Most of the researchers, on the other hand, understood that they couldn't expect ever-increasing transfers from the State to the local economy, especially with the world's trends toward free trade liberalism and the declining role of the State. They felt that priority should be given to developing autonomy and self-development through good economic management and accountability toward the funding agencies—public or private. These two concepts often clashed, but the positions were rarely discussed openly. MPST directors preferred to rhetorically support accountability and self-development, while in practice accepting flexible management from their members' organizations and practicing it themselves.

Environmental Concerns

In the end, LAET researchers concluded that it was doubtful that the farmers had any interest in natural resources and ecology as such. Many authors believe that farmers worldwide have no interest in environment per se and that their first concern is the continuity of their farm and their family (McC Netting 1993). In the frontier, attachment to the land as the basis of the family inheritance is especially weak. Some authors doubt that frontier farmers are really custodians of the land, as they are sometimes portrayed. The strategy of frontier farmers seems to be characterized more by the unrestricted conversion of natural resources (forest and land) into money and degraded pastures in a strategy of mining (Lena 1986; Schneider 1995). Some farmers do not hesitate to declare that they have no qualms about quick deforestation of the region and about land concentration in the hands of a few owners, if this brings them monetary profit. Other, supposedly environmental, groups may also be primarily concerned with short-term benefits. Various authors have observed that adopting ecological themes by rubber tappers' unions and Indian federations seems to be aimed at an alliance with foreign NGOs and access to international funding rather than a genuine concern for the rainforest (Redford and Maclean Stearman 1993; Geffray 1995).

Convergence of Environmental and Social Concerns

In contrast to the seeming indifference of many farmers to sustainability, it became clear during strategic planning meetings with LAET that MPST's desire for stabilization of family agriculture for future generations meant that it must plan for management of natural resources at the regional level. The leaders of MPST began to understand that massive deforestation would harm the family farm through an increase in fires and invasion of weeds. In addition, conversion of many small farms into large-scale pastures would result in the concentration of land in medium- and large-sized ranches. Such a trend would mean the end of farmers' unions, since their base would disappear and probably be replaced by ranchers, who would join the competing patrons' unions. Another factor was the never-ending extension of feeder roads farther and farther from the main road. Such proliferation would make road maintenance and the provision of social facilities more difficult toward the end of the feeder roads, while facilitating land concentration close to the main road. Finally, the rapid exploitation and subsequent exhaustion of commercial timber in the region could bring an artificial economic boom of short duration, followed by a long-term depression. According to an MPST leader, such a short-term boom would probably reinforce the political enemies of the MPST more than their friends.

In conclusion, we can say that the farmers'organization had some real interest in the better management of natural resources in the region, not because of ecological consciousness, but the realization that natural resource management was one of the factors affecting the future of family farming in the region.

Failure to Communicate?

The fact that LAET did not succeed in establishing a common strategy with MPST was not because MPST had no real interest in a better management of natural resources. Actually, MPST's understanding of better management was different from that of the LAET researchers. One needs to consider that this interest in natural resource management was just one aspect of a much broader set of MPST objectives. These results suggest a failure in the constructivist model to establish a common strategy through improved communication.

The Researchers' Perspective

As with the farmers' organization, the researchers' priorities and activities often did not correspond to their official objectives. On many occasions, the researchers affirmed publicly that they were fully committed to the development of family farmers and were doing their best to assist them both technically and politically, and to reinforce their organizations. However, in practice, they sometimes showed little interest in the farmers' concerns. In some cases, the researchers were reluctant to volunteer for activities that implied regular contacts with farmers, either through applied research or direct assistance. For example, a farmers' group to which LAET had promised regular follow-up remained without any visit for many months, because the researcher who was responsible always had other priorities. Part of this reluctance can be attributed to the lack of confidence of young graduates, who are hesitant to enter into direct technical assistance, because errors at this stage are quickly publicized. However, it may also have come from the conviction that research has a higher status than simple extension and that, for career development, it is much better to be recognized as a researcher than as a good extension agent. Objectively, this is certainly correct in general, particularly in the Brazilian context. Young agronomists interested in continuing in the field of small farmer development (assuming that they had already made a choice *not* to work for the big estates or agro-industries) can get a job from the NGOs, often of a precarious and insecure nature, from the university, or from a federal institution. In the latter case, public competition and future promotion are largely based on diplomas and publications. And, obviously, a candidate's experience and success in the field of development have no value. Another possibility would be to work for the State Extension Service (EMATER), but they are so badly paid and demoralized that it hardly constitutes an alternative.

A second problem from the researchers' point of view was that the young researchers felt they were already making a big sacrifice by working in Altamira, and they usually tried to spend their weekends in the city. The older researchers did the same for family reasons. This seems a reasonable attitude, but it also seriously limits the time spent in the field and in the communities, considering that it takes a full day to reach the farmers' communities, even with a good truck. As a result, development-orientated activities often suffered due to insufficient involvement from the individual researchers.

A third problem stemmed from the fact that researchers were occasionally involved in political activities. Some were members of political parties and could enter into local party politics. Because of their higher level of education, they enjoyed a special status that they could use to run for responsible positions. One researcher was offered a job as agriculture secretary of Altamira District. Another was elected coordinator of the university at Altamira. Because the previous coordinator had been a mayor of Altamira, some MPST militants considered the position of agriculture secretary as a start for a local political career.

Most problematic was that various LAET researchers were suspected of belonging to an internal faction of the Partido dos Trabalhadores, opposed to the one to which most MPST leaders belonged. They were therefore suspected of organizing internal opposition within the organization. Whether this was true or not, such an accusation was serious enough to provoke a strong reaction from MPST. Again, it must be seen that researchers with ambitions for university positions realize that the way to obtain responsible positions, such as department head or presidential positions, is to link with a given political group, since all positions are elected by the staff and students together. Considerable time and energy were spent in these strategies and internal conflicts at the expense of development activities and action research in the farmers' interest.

Conflicts and Competition of the Researchers with the Farmers' Leaders

The political behavior of some researchers sometimes provoked the MPST to see them as possible political competitors who first manipulate local leaders while pretending to advise them, and then use their relations and prestige to gain positions in local politics. MPST leaders were especially sensitive to this risk and said at various times that they wanted to keep the "hegemony" of farmers in their own organizations. One of the measures taken to avoid possible interference was to limit, as much as possible, direct speech of the researchers during assemblies and meetings of the farmers' organization and even to avoid inviting them to these assemblies. The wisdom of these measures was later confirmed when some LAET researchers developed personal ambitions in the region.

Some of the local leaders (and MPST directors themselves) were unwilling to recognize the role played by LAET, which often ran contrary to their own strategies and could be seen as a threat to their leadership. For example, criticisms and suggestions about ill-managed projects were seen as criticisms of the leaders and were not accepted in public. The use of agricultural credits is another point of contention. In these conflict situations, local leaders preferred to downplay the role of LAET, probably to avoid future criticisms in their own organization.

Lack of Confidence and Transparency

The main priorities of MPST were not initially clear to the LAET team. For example, when LAET observed that all activities were suspended during six months for the 1993 state election, so that MPST could support their outgoing coordinator as a candidate, researchers wondered whether MPST's main objective was not simply political power, with other activities organized more to attract the people than to solve their problems. It was difficult to have a clear discussion of this with MPST staff, because they had a culture of dissimulating their true strategies. In addition, they had their own doubts about LAET's motives and strategy. Later, MPST leaders admitted that they could not show their strategy to organizations who might divulge it to "enemies." In part, such secretiveness might be a component of the Brazilian culture in general and is probably more pronounced in frontier farmers, owing to their domination by ranchers and the government. Furthermore, MPST evolved from a semiclandestine tradition of groups under a dictatorship. Their main success in the 1980s was the takeover of the existing unions, previously controlled by conservative syndicalism.

Even within MPST, there is a constant mistrust of new leaders, who might use their positions for personal benefits and sell themselves to the local elite factions. This mistrust was obviously even more pronounced when intellectuals from outside the region announced that they would help the farmers' organizations. Historically, this mistrust was fully justified by the numerous cases in which local people had been lured and betrayed by "nice-talking" intellectuals, who later forgot their promises and used the votes and support of the locals to climb the political ladder to Brasilia. Indeed, the local history of political paternalism is full of these events.

LAET, as a group, was more transparent in its objectives and strategies to MPST. However, its individual members also had hidden personal strategies that were perceived by MPST, which reduced the trust in LAET's capacity to keep its promises in the long term.

Communication Problems

LAET researchers believed in 1995–1996 that they had agreed on a long-term strategy with MPST. LAET researchers thought that when MPST was invited to a meeting, the persons who participated represented the whole organization. This later proved to be false, due to (1) the limited capacity and interest of the individuals representing MPST to discuss the content of these LAET-MPST discussions with a wider group, (2) the lack of custom to record the discussions or read the minutes that usually were written by LAET, and (3) the habit of discussing important issues with historical leaders of MPST who were no longer in the region, which took much time and was not always conclusive. Another important factor is the cultural fact that farmers were not accustomed to directly criticizing or confronting other persons in public and preferred to use indirect forms to "send a warning" when they disagreed. Many times, the lack of reaction should have been seen as a disagreement with what was being said or proposed (or at least a lack of clarity about the proposal and its implications) and not as an indirect agreement. In this cultural environment, one should not assume that "who does not speak, agrees," but rather "who does not reply, disagrees."

Diversity and Conflicts Within MPST

Only four years after the PAET program began, it became clear to LAET that there were various views and individual strategies within the MPST itself and that its strategy could not be described as unambiguous. Like most voluntary organizations, decisions were the outgrowth of complementary and sometimes conflicting views from other "sister" organizations (namely, the Catholic Church, the Labor Party, the unions, and other NGOs), from historical founders of MPST, as well as present directors. It was actually naïve of LAET to think that MPST could have a single strategy. However, only in the time of crisis did these contradictions become apparent.

Most popular organizations in Brazil face two conflicting visions of the future: an elitist Leninist vision of a conquest by enlightened intellectuals who will later help the masses and an anarchist-syndicalist vision giving priority to education of the masses ("popular education or conscientization") as a necessary step preceding any political change. This latter line was developed mainly by radical Catholics (Freire 1970) and the "theology of liberation," although the church was not devoid of contradictions when applying it. It is not clear whether MPST was able to present a coherent long-term strategy, given the lack of agreement as to which strategy would be most beneficial and workable. The directors probably prefer to maintain a deliberate fuzziness regarding MPST's ultimate objectives and strategy to avoid fueling an internal debate, which might result in fracturing the organization.

Toward a Better Understanding

Necessary Humility

Large private corporations usually have specific "rules and policies" regarding conflicts and censorship on restricted information. In the local farmers' organizations, things are not so clearly spelled out. In the case of conflict, the farmers' usual defense is to close themselves off and to not continue the debate. This can be seen as a typical trait of peasant farmers around the world (Hebette 1996). The researcher has to make a serious effort to "feel" the limits (unwritten rules) and what is expected. Most LAET members lacked experience in this field and probably often took positions that were considered offensive by the farmers' leaders. Humility and discretion are essential in all cases to avoid conflicts.

Setting Limits

The tumultuous history of the relationship between LAET and MPST taught the researchers, and probably the farmers' leaders, several hard lessons. One was that each group should clearly establish the areas that are critical for its future, which should not be interfered with by the other group. For example, for LAET researchers, as for researchers in general, publication is a critical area. But to publish and demonstrate innovative development results, they have to be able to carry out trials over a long period of time to be sure of their results or to fine-tune the methodology. The leaders of the farmers' organization often created problems for the researchers by instilling expectations in the farmers that immediate benefits would be forthcoming.

For the MPST, an important priority for the leaders is to show their authority and competence to rank-and-file members. To help them do this, researchers should avoid taking positions that undermine such authority, even though researchers believe that, on certain issues, the leaders may be wrong or misinformed. This may be contrary to the researchers' ethical view about their role in society (that they should always speak the truth in public, no matter what interests might be affected). However, it is a realistic condition that researchers must accept if they hope to work durably with local organizations in the name of a wider ethical involvement.

Evaluation of the Partnership Between Researchers and Farmers

Evaluating the Initial Assumptions

In Chapter 5, we presented three assumptions about the role of farmers' organizations such as MPST in natural resource management programs. The initial assumption—that farmers' organizations have *some* interest in better management of natural resources—was confirmed. However, this interest was not based on the usual ecological values, but on a perception of small farmers' interests as a group in the medium and long term. Organization leaders were more conscious of these long-term collective interests than the lay members of the organizations. In addition, this objective was only one of various objectives, which include political results, short-term improvements for the farmers, and their own organizations' reinforcement. These various objectives are often contradictory and sometimes create conflicts within the organizations themselves.

The second assumption—that farmers' organizations can help in the participatory actions program—was partially confirmed in that the farmers' organization:

- Maintained strong pressure for client-oriented results
- Disseminated results, at least when results corresponded to their interests (e.g., they quickly diffused the idea of the Casa Familiar Rural as well as the use of leguminous cover crops)
- Facilitated the research in certain stages, especially at the beginning
- Represented the farmers in negotiations with the Bank and Extension Service
- Effectively acted as a link between field-level activities and policies at the regional and national levels, especially in credit, marketing, and land policy

However, the MPST also sometimes caused difficulties or even blocked research, as shown in the Uruará case study in Chapter 6. The organization did not always reflect the farmers' immediate interests and demands, and often concentrated on the organization's and the leaders' personal interests. Considerable effort and time were invested by LAET in building a relationship with the efforts of MPST, time that might have been more immediately useful in direct work with farmers. It is, therefore, difficult to say whether the net effect of working with the farmers' organization (MPST) was positive or negative in terms of the efficiency of PAR. Partnership certainly did not automatically result in the expected benefits.

The third assumption was that an effective common strategy could be reached. However, despite the time and effort consumed by meetings, and despite the fact that the representatives of the MPST and researchers did agree on a common analysis of the environmental questions, it was not possible for the two organizations to reach an effective common strategy. The conflict that erupted between LAET and MPST showed that satisfactory communication between the two organizations had not been achieved.

Reformulating the Model

What was wrong with the assumption that it would be possible to define a common strategy? As a result of the PAR experience, some answers have emerged.

The question of sustainable development was only one of many objectives of the farmers' organization. Therefore, its overall strategy was very broad and contained conflicts between various objectives (particularly political and development objectives). The research team also had to arbitrate conflicts between its scientific and institutional objectives and its development priorities. *A common strategy, therefore, cannot be restricted only to elements of common interest.* Each partner also needs to understand all of the other partners' objectives and strategy to negotiate a common program that will maximize the positive results of the cooperation and minimize its negative side effects.

Another incorrect assumption of the model was that farmers' organizations, being strongly structured, could formulate and present a unique long-term strategy. In fact, due to the existence of various factions within MPST, it was not able to formulate such a strategy. The researchers also had diverse personal interests and professional strategies, which were not always transparent within the research team itself. This became clear during moments of crisis, in which the fiction of "unity" of the research team was shattered. The research action strategy therefore should incorporate this uncertainty.

A good, common understanding can be achieved only if each partner agrees on *the rule of transparency,* that is, completely and honestly disclosing agendas that culturally are often kept secret. This was definitely not done by the MPST. Dissimulating strategic information about its own objectives and functioning was clearly part of its organizational and political culture. This tendency, in turn, was reinforced by a serious lack of trust in the objectives of LAET. However, distrust was justified historically by the traditional posture of technicians and intellectuals in Brazil, to use their prestige to get into power positions. Distrust was compounded by nationalist views about international intervention and interests in Amazonia. Finally, on various occasions, MPST tried to use the research organization (especially in Uruará), rather than to try to negotiate an acceptable compromise.

To a lesser extent, LAET can also be accused of having used MPST, for example, when LAET obtained MPST support for the licenciatura (degree program) in Altamira, based on the hope that it would prepare technicians for local organizations. In fact, it was mainly a response to bureaucratic requirements of the local university. LAET also can be rightly accused of having used MPST support to conduct research that was primarily academic.

As time passed, the manipulation and dissimulation strategies lost their efficiency. Researchers naturally started to question the cooperation when a contradiction between certain "official objectives" and effective practices of MPST became obvious. This questioning led to increased conflicts, countermeasures to avoid future manipulation, and conscious efforts to better understand the MPST hidden agenda. Reciprocally, MPST questioned the distance between LAET's official objectives and its real priorities and results. This finally led to a major crisis, which helped to clarify even more of each other's real position. One can safely assume that if a new agreement is negotiated after this crisis, it will have a more realistic basis and content than the first one. On the other hand, if not controlled, a conflict can lead to increasing retaliations and end up in complete opposition and separation.

The conclusion regarding the model of interaction between farmers' organizations and researchers is that a constructivist model, based only on the concepts of knowledge systems and individual (or collective) strategies and objectives, is insufficient to analyze the interaction among different actors within a research action program. The local actors can and often do use strategies of dissimulation and manipulation of information, which interfere in the negotiation and communication process. The same practices have been found to be widespread in many development programs (Olivier De Sardan 1995).

Relations between organizations are always influenced by power struggles. Knowledge is one source of power. Political organizations are especially conscious of this fact and are inclined to manipulate information to their advantage. The same can be said of research and academic organizations, even if certain "professional standards" are supposed to restrict information distortion by researchers (see Latour and Woolgar [1986] for an account of sometimes unethical research strategies). Neglecting to analyze this political aspect of relations between organizations and individuals results in a limited understanding of the process of social change. This limitation is particularly relevant for the type of research action programs that are discussed here.

Methodological Recommendations

Research groups that seek to develop long-term agreements with local organizations should incorporate or associate on their team a trained anthropologist, someone who is familiar with the local culture or a mediator who could be sufficiently familiar with both universes, such as a young farmer who had the opportunity to study or an intellectual who accompanied the creation of farmers' organizations. Regardless of the mediator, the question of his or her loyalties must be addressed.

In farmers' organizations and other local groups, special attention must be given to an analysis of their critical arenas. What are the sources of the organization's power? How important is the organization to the State and to the other local organizations and actors? More specifically, what is the source of legitimacy of the elected leaders within the organization? If the researchers do not have a clear idea of these critical arenas, the chance is that they will involuntarily and possibly unconsciously compete with the local leaders for recognition and legitimacy. Such competition will occur in arenas in which local leaders are most sensitive: in the discussions and contacts with base members of the organizations and in negotiations with the State and other funding agencies.

The researchers, too, have to make explicit their own critical arenas early in the game: scientific recognition and concrete and replicable results in the field, but not necessarily on a large scale. If farmers do not recognize this need, they may unwillingly interfere with the small-scale experiment that researchers have patiently planned and elaborated.

Implicit or explicit rules concerning the behavior of the partners within these critical arenas should be devised as early as possible. This should help in building trust and facilitating open communication. Better communication should, in turn, help partners to understand internal contradictions and conflicts in each group earlier and how to get along with them. It must be borne in mind that time is essential in establishing such communication and that learning to cooperate is a long process.

Indicator of Success

The best indicator of our understanding of other actors is that we finally manage to comprehend them well enough to be able to predict their next reaction to given circumstances with reasonable success, and to consequently develop meaningful discourse in which misunderstandings and surprises are minimized and subsequent cooperation maximized. This is tested during the participatory action research process.

Deforestation in the Brazilian Amazon: A Comparison of Conventional Diagnoses and Diagnoses Based on PAR

In this chapter, we compare the conventional understanding of causes of deforestation with the understanding obtained by PAET (Programa Agro-Ecologico da Transamazônica), through the participatory action research (PAR) approach. By "conventional," we mean an understanding based on a review of the technical and scientific literature, not merely popular articles or newspaper reports.

Most conventional diagnoses of deforestation in the Amazon have not been tested. Analyses have been made and hypotheses have been set forth, but no effort has been made to obtain feedback that would determine the validity of the hypotheses. In contrast, feedback is a fundamental aspect of PAR. As a result of feedback, diagnosis of the problem often changes and actions based on the initial diagnosis are modified to accommodate the new findings. After the experiences in Uruará, Porto de Moz, and Altamira, it became clear to Laboratorio Agro Ecologico da Transmazônica (LAET) researchers that their initial diagnoses were insufficient. After the initial LAET diagnosis, a series of evaluations and revisions allowed researchers and participants alike to gradually work their way toward a better understanding of the problem and to move more effectively toward its solution.

This chapter begins with two conventional diagnoses of the social, political, and economic dynamics that led to the recent increase in colonization and deforestation in the Amazon region. The second part of this chapter presents the LAET diagnosis after five years of action research and compares it with conventional diagnoses.

Conventional Diagnoses

Deforestation in the Amazon has been explained by several different (but not necessarily exclusive) models:

- The *government incentive model*. Government incentives have played a crucial role in the frontier expansion since the 1950s.
- The *natural expansion model*. Frontier expansion in the Amazon is the continuation of a long-term trend of land occupation in Brazil, a process occurring almost naturally, given the demographic and economic expansion of the country.
- The soil exhaustion model. The low soil fertility in most of Amazonia requires settlers to continually abandon plots that are exhausted and to move to clear new areas with virgin soils.

The Government Incentive Model

Many authors agree that the policy of the federal government has had a significant impact on the occupation of Transamazônica since the 1970s. Successive military presidents declared a need to "integrate and not to give away" (integrar para não intregar) the Brazilian Amazon. The Amazon was viewed as a rich mineral province that might be claimed by foreign governments or become an international territory if Brazilians did not physically, economically, and militarily occupy the region. There was, and also still is, a concern about regional inequalities that could lead to separatism or local revolutionary guerrillas. Finally, it was felt that the immense Amazonian Basin could absorb the excess rural poor and landless from other regions, especially the northeast. It would therefore help to solve the national land distribution problem without resorting to land reform that would be opposed by the traditional elite.

Another factor influencing the government was the opinion of certain influential US-based scientists who claimed that, with the adoption of new technologies, especially chemical fertilization and liming, tropical lands could become the most productive in the world (Abruña et al. 1964; Sanchez 1976; Nicholaides et al. 1982). As a result of these claims, many people became convinced that Amazonia could transform itself into the biggest granary of the world, and the Brazilian agricultural research agency (EMBRAPA) devoted much effort trying to fulfill this fantasy.

The first step in implementing the policy of Amazonian development was for the government to open an extensive network of roads in Amazonia. The first was the Transamazonian Highway, opened in 1971 between Marabá and Humaitá. The second was BR 364 from Cuiabá-Porto Velho in Rondônia. These roads were used to implement the colonization program that took place between 1972 and 1975. Small landowners or those who were landless from southern and northeastern Brazil were recruited, transported, provisionally housed, and given a plot of virgin forest together with a small stipend for the first year. The program turned out to be very costly and not as efficient as planned. Abandonment after the second or third year was common (Moran 1981). After 1976, the government went back to its previous policy of giving priority to extensive livestock ranches. This policy provided tax and credit incentives to companies or businessmen who wanted to invest in Amazonia and gave them priority in land distribution. However, once the spontaneous migration of farmers from the south and northeast had been initiated, it did not stop. As a result, many conflicts and much violence broke out between farmers and ranchers, especially in southern Pará.

Government policy also had negative consequences in Acre, where speculators and ranchers confronted traditional extractivist seringueiros, who had used the intact forest sustainably for generations.

Large Ranchers

Hecht (1984) concluded that most land occupation that occurred under the government's policy of fiscal incentives was motivated by speculation and fiscal incentives, rather than by actual agricultural or beef production. Practically all the big estates engaged in cattle production, but Hecht (1993) observed that the yield and profitability of this production was extremely low due to poor management, which led to rapid degradation of pastures and poor sanitary condition in the herd. She concluded that large-scale conversion of forest to pasture was a strategy to secure land titles and to prevent land occupation by settlers. To secure a land title under Brazilian law, it is necessary to prove that one has "improved" the land. This is done most easily by clearing the forest, putting in a wire fence, and allowing a few cattle to graze. The cheap, subsidized credit that then becomes available could be reinvested in more profitable business operations.

Sawyer (1990) also concluded that a large part of Amazonian deforestation after the mid-1960s could be explained primarily by government policy, which benefited a limited number of industrial and financial groups. However, by the end of the 1980s, large ranching operations, and the fiscal and credit incentives extended to them, could explain only a relatively limited part of actual conversion of forest to pasture (around 20% in the Pará region according to Mahar 1990). The bigger growth in cattle herds was observed in small holdings, which did not benefit from these incentives.

Small Farmers

Large ranchers were not the only ones engaged in speculation. Some small farmers did so as well. These farmers frequently converted forest into pasture and then sold the land to ranchers (land under pasture is estimated to be two to three times more expensive than land under forest). However, other small landowners were genuinely interested in cattle production. De Reynal et al. (1995) observed that, in Marabá, cattle productivity on small farms was relatively high (65 kg of live weight gain per hectare per year in the initial stages of pasture development), leading to good labor returns ($4.00 to $8.00 per work day) because of low labor requirements. In contrast, returns from rice production were lower and more variable (from $1.00 to $8.00 per day). Rice grows well only in areas with good natural fertility and is profitable only where farms are close to cities.

Pasture Degradation

Pasture degradation is generally considered to be almost inevitable. After 6 to 10 years, the carrying capacity drops from 0.5 to 0.1 adult animal per hectare, whereas the work needed to clean the pasture increases sharply. As a result, pastures are usually abandoned and new land is sought. When no more forest is available to convert to pasture, the system enters a period of crisis. In the 50-hectare plots typical of the small producer, this usually occurs after 10 to 15 years. At this point, the farmer has the choice of either selling some of his cattle or selling the lot and most of his cattle and buying new land in the forest farther from the main road. Annual or perennial crops are no longer an option, since increased weed pressure and risk of fire make their production unviable. As a result, many authors (e.g., Fearnside 1990a) have concluded that, in both economic and ecological terms, the policy of fiscal incentives to encourage ranching was not a viable enterprise. However, researchers from an independently funded non-government organization (NGO; IMAZON [Instituto do Meio Ambiente Amazonico, Belém]) found that, much to their surprise, cattle production near growing centers of population can be economically viable as a result of reduced costs of transporting cattle, improved markets, and lower costs of factors such as fertilizers that artificially maintained pasture fertility.

The effect of forest conversion to pasture on soil fertility has been a controversial subject. Some EMBRAPA researchers claimed that pasture improved the soil's fertility (Falesi 1976). But various authors observed that pastures were quickly degraded and abandoned and that soil fertility could not be maintained under natural pasture (Fearnside 1980). Serrão et al. (1978) identified the decline in available phosphorus as the main factor responsible for declining yields. However, other factors, such as increased soil density and invasion by shrubby weeds, were also observed to cause pasture degradation (Hecht 1984). Jordan (1989) reported that nutrients from decomposing logs actually do enrich the soil for two or three years, but soon thereafter the nutrients are leached, volatilized, or fixed in an unavailable state, causing productivity of grass or crops on the deforested land to decline.

Recovery of degraded and abandoned pastures into forests is inhibited partly by predation from leaf-cutter ants and competition from vigorous weeds (Nepstad et al. 1990). An increase in the frequency and area of fires, together with the opening of the landscape, may also play an important role in the "setting back" of succession in Amazonia (Uhl et al. 1988) as well as in other regions where forest was converted to pasture (Hopkins and Graham 1984).

The Frontier Expansion Model

The second model to explain deforestation in Amazonian Brazil is the natural expansion of the frontier. Defenders of this model note that the processes of occupation and deforestation of other states have gone through the same pattern. For example, Paraná lost 95% of its forest cover between 1940 and 1970 through natural expansion of the frontier (Lena 1986).

Coy (1996) describes the natural expansion process as a series of three phases. During the first period, poor migrants or adventurers occupy remote lands away from government authority and practice diversified subsistence agriculture. In the second phase, infrastructures are established and land tenure becomes more secure. Land becomes concentrated in the hands of fewer owners and extensive livestock production takes over. During this second phase, important social differentiation occurs. Some farmers manage to increase their capital and land, whereas others are forced to sell their land and may have to work as day laborers. In some cases, the second stage differs somewhat, when farmers develop perennial crops rather than extensive livestock, as in Rondônia (Leite and Furley 1985) or the Transamazônica (Hamelin 1990). In these cases, land concentration is not so pronounced. In the third stage, as the region becomes fully integrated into the national transportation network and economy, land prices continue to rise and extensive livestock production is replaced by modern crop production, which gives higher returns per hectare but needs higher investments.

Various authors consider that, during the first phase, small farmers are actually adopting a mining-and-migrating strategy and that they are the main agents responsible for the deforestation in Amazonia (Myers 1984). De Reynal's modeling of farmers' accumulation strategies in Marabá confirms the existence of such a mining strategy among some small farmers (De Reynal et al. 1995). Mahar (1990) also considers that present natural resource exploitation in Amazonia can be characterized as "mining." He defines this as both wood extraction (logging) and unsustainable agriculture, because they are practiced in a way that does not permit the recuperation of the forest or soil. He estimated that, in approximately 10 to 20 years, a complete cycle of logging, annual cropping, and ranching leaves the soil completely exhausted, at which point the land is abandoned. Mahar states that this mining cycle continues as long as land remains relatively abundant and cheap.

From the economic point of view of individual farmers and owners of timber industry, it is actually rational to "mine" as long as land is cheap, compared with the costs of sustainable management. For those who get into the mining cycle early, benefits are substantial. Because they don't want to lose their advantages, they give little political support to any policy that reduces free land access or promotes sustainable agroforestry. This lack of support, together with the physical size of the area to be controlled, results in a very high cost of integrating Amazonia into the national legal infrastructure, a cost the federal government is not prepared to pay. Besides the financial cost, Mahar (1990) suggests that there may also be a political cost to closing the frontier, since the myth that the frontier represents opportunity for those with ambition and courage probably continues to defuse pressure for redistribution of wealth and income in the country.

Was Frontier Colonization a Failure?

Many authors have claimed that the agricultural colonization of Amazonia was a partial or complete failure, based on early observations of high rates of abandonment in newly colonized areas. Moran (1989) related abandonment to lack of knowledge of the local ecosystems by newly arriving migrants. Fearnside (1991) stated that colonist agriculture was bound to fail, since population pressure and lack of adapted agricultural traditions lead to insufficient fallow periods. However, Schneider (1995) observed that, for many colonists on the frontier, the economic situation actually improved. An INCRA (Instituto Nacional da Colonização e Reforma Agraria) survey of 46 public agricultural settlements, including

10 settlements in the North, showed that on average, colonists in Amazonia obtained a better income than they had in their regions of origin. Since many were previously landless, becoming landowners always meant an improvement in both economic and social terms. Colonists accumulated capital assets at a fairly high rate of 18% per year, higher than in other regions except in the south of Brazil. However, the rate of departure of colonists from the North was much higher than in other regions (between 20% and 60% of original settlers). Many of those who left sold when land prices rose. This suggests that the reason for their leaving was a "mobile accumulation" strategy (a search for increased wealth) and not economic failure.

Schneider (1995) gives a deeper analysis of economic rationality by introducing the factor of a high discount rate. Since Brazil has one of the highest rates of interest in the world (around 30% per year), extremely unsustainable practices giving high returns at short terms have been promoted. Within the classic theory of economic choices, Schneider shows that this has encouraged unsustainable and mining practices of agriculture, livestock production, and forestry. He explains that small farmers selling out to capitalists, observed during the second phases of the frontier (e.g., consolidation of government presence, property guaranteed), occurs as a result of different net production values. The frontier colonists have little access to capital and technology; consequently, their productivity is low compared with that of the capitalists. In addition, frontier colonists are not accustomed to dealing with the bureaucracy that accompanies the integration of a region into the national infrastructure.

Although both the government-sponsored and volunteer colonization programs did benefit some of the early colonists, the process generated little income for the country. This was due to the low level of agricultural yields and the high cost of public investment, such as road-building and sacrifice of public forests and reserves. Amazonian agriculture represents only 0.5% of Brazil's gross domestic product.

The Soil Exhaustion Model

Many authors have hypothesized that soil-fertility decline is the main cause of decreasing yields and failure of agricultural colonization in Amazonia (Fearnside 1982; Jordan 1987; Eden 1987). Although burning of vegetation increases the pH of the soil, increases phosphorus availability, and reduces the effect of aluminum toxicity, these effects are short-lived. The quantity of organic matter remaining in the soil is the critical element for sustained yield, particularly because of its influence on phosphorus availability. Two or three years after clearing and burning, the organic matter on and in the soil may be nearly depleted. This is when production declines (Jordan 1989).

Most authors suggest that slash and burn is sustainable if the fallow period between cycles of cultivation is sufficient. The fallow period must be long enough to restore sufficient nutrients and organic matter to the soil to support several years of cropping. Characteristics of the fallow vegetation and soil are important, but longer periods of fallow usually result in higher concentrations of nutrients and organic matter, which in turn can support longer periods of cropping. With a given level of soil nutrients and organic matter, the length of the cropping period can vary, depending on the crop planted and the type of cultivation used.

The sustainability of farming systems also depends on landscape factors, such as the following:

- *Seed rain* and, therefore, the fallow composition (Brokaw 1985; Nepstad et al. 1990)
- *Local climate* and fire risk (Bushbacher et al. 1988)
- *Perennial weeds* in cultivated plots (Scott 1987; Mt Pleasant 1990)

Recent research in two different sites in Bolivia (Wilkins 1991; Staver 1991) and in Peru (Sanchez, Buschbauher, Uhl, and Serrão 1987) suggests that increasing weed pressure may be a more important factor than soil fertility per se. Weed competition is severe in low-fertility soils, since weeds are better competitors than most crops for the limited supply of nutrients. Contrary to most expectations, Thill (1991) in Bolivia and De Reynal (1995) in Marabá, Brazil found that yields of rice after the clearing of a short fallow period (five years' secondary regrowth) were not significantly lower than after the primary forest. This again suggests that nutrients were not the factor—the fallow periods were just long enough to decrease the pool of weeds. However, after repeated cycles, reestablishment of nutrient stocks can eventually become the critical parameter that controls minimum time of fallow. In regions with sandy, acidic soils, nutrients become limiting more quickly. De Reynal et al. (1995) observed that second fallows in Marabá needed to be long (10 years) for secondary vegetation to reach the same stage and suppress weeds as effectively as the first fallow (five years).

Based on these different observations, slash-and-burn agriculture might be sustainable on most Amazonian soils with a 10- to 15-year fallow/2-year cropping cycle. If this is true, and considering that small farmers deforest an average of 3 hectares per year, slash-and-burn farming should be sustainable on 30-hectare farm lots, an area much smaller than the average lot given farmers in most settlement programs, in which plots range from 50 to 100 hectares. With 30-hectare lots, there would be a density of 3.3 farmers' families per km^2 or approximately 20 habitants/km^2.

As long as population densities on the frontier remain below that level, slash-and-burn agriculture can be sustainable. This analysis is also corroborated by Ozorio de Almeida (1992), who showed that annual crop yields have substantially increased, rather than decreased, on a regional basis as population increased. However, the idea that plots in frontier areas can be continuously farmed through conventional means with chemical fertilization and liming (Sanchez 1987) appears unrealistic, because it involves continuous monitoring of soil status and complex nutrient input, which is costly and unavailable to farmers.

Proposals to Mitigate Deforestation

Some authors consider population growth to be the fundamental factor in expansion of agriculture into forested lands (Moran 1996). However, the low human density in Brazil tends to refute this hypothesis (Anderson 1990a). In fact, the number of migrants to Amazonia is relatively small compared with the total number of persons who migrated from rural to urban environments (1 million out of 20 million in the 1970s and 1980s). On the other hand, there is general agreement that the very unequal land and income distribution, especially in Northeast Brazil, has been the major explanation for the migration of landless farmers to Amazonia. Given that these poorly educated migrant populations are not likely to be absorbed by the urban economy for several generations and that agriculture will remain important in Amazonia for some time to come, countering deforestation will require initiatives to help Amazonian agriculture become sustainable.

Fearnside (1991) has always been a vigorous opponent of agricultural colonization in Amazonia. He suggested canceling all policies promoting migration toward Amazonia, especially that of building new roads in the region. Another policy that would reduce migration to the Amazon is agrarian reform in other regions of Brazil. Such a measure would reduce social disparities in these regions and lessen the incentive for the disadvantaged to leave. For colonists who are already established, Fearnside suggests that agriculture should be oriented toward more intensive and diversified agroforestry systems. However, he recognizes that most perennial crops suffer from a limited market; therefore, one should not be too optimistic about the potential of these crops in Amazonia.

Fearnside (1986) has also called for a general zoning of Amazonia that would separate areas reserved for low-impact forestry exploitation, extractivist activities, and limited agriculture. However, he recognizes that the possibility of actual implementation of zoning in Amazonia is remote, since the law in most of the region is little respected.

In contrast to Fearnside, Hecht's (1984) opinion was that the impact of small farmers on Amazonian ecosystems was limited and that large-scale ranching is mainly responsible for deforestation and environmental degradation. Since she considered that tax incentives, subsidized credit, and land speculation were the main factors responsible for this trend, she concentrated her recommendations on changing these policies.

Mahar (1990) suggested that the government should suppress incentives to ranching and should instead consider subsidizing small, low-impact agriculture, including agroforestry, extractivism, and small animals. He also suggested that the government enforce a new land-tax policy, including a general land tax that would discourage extensive ranching or forest mining, and would enforce the taxation of profits derived from land sales (25% in theory) to reduce the incentive for land speculation. Moran (1996) suggests that logging activities should also be monitored more closely and recommended stronger research on intensive agriculture with low external inputs. Schneider (1995) goes further in his proposals. He suggests that the provision of cheap credit and easy titling for small farmers during the first phases of land occupation would result in a quick increase in land value. This would reduce the incentive for urban-based entrepreneurs to buy this land and would encourage agricultural stabilization. On the national level, Schneider suggests increasing job opportunities and reducing social disparities in other regions, as well as reducing interest rates overall. Locally, he suggests that providing good-quality social services (especially health and education) and infrastructure (good transportation) would also stabilize small farmer settlement.

Feasibility of the Proposals

Would any of these proposals do any good? Sawyer (1990) is pessimistic. He believes, "If deforestation has profound social causes, there is no easy technological solution or 'fix,' neither can nature be protected by Decree" (p. 266). However, he observes that most of the initial pull for occupation came from a government policy that was mostly concerned with national security. In terms of the national economy, this proved to be unfavorable. Only restricted groups within the capitalist sector, such as the local elite, have profited. However, this elite had only a limited weight in the national Congress. There was no deep support from national business interests to continue encouraging colonization in the Amazon. As a result, public investments in Amazonia have continuously decreased in the 1990s. Sawyer concludes that, in this case, there is no contradiction between environmental concerns and democracy: there is room for an alliance among social movements, national government, and international agencies around a more equitable and sustainable model of development. This has actually been demonstrated by the rubber tappers' movement in Acre (Allegretti 1990).

In a more detailed analysis, Schneider (1995) discusses which measures could receive sufficient political support, locally or nationally, to have a chance of success. He considers, with good reason, that national policies without local support are bound to fail. For example, although the taxation of land sales proposed by Mahar (1990) would be sound from an economic and political point of view, it is doubtful that it could be enforced efficiently, considering the high level of tax evasion, corruption, and lack of control of land tenure found in Amazonia.

However, Schneider (1995) distinguishes between support from the poor majority and support from the local elite. Giving small holders priority to land claims would receive support from the local poor, but would probably be opposed by some of the local elite. Only with appropriate federal support could this plan have some chance of success. Such support could be in the form of an association among the federal government, local farmers' groups, and NGOs. In contrast, restrictions on uncontrolled logging are politically unrealistic, because they would be opposed by both the rural poor and the local elite.

In Schneider's (1995) opinion, zoning is also likely to be rejected by everyone locally, since everybody has, or might have, some interest in unrestricted access to new lands. It is well known but little discussed in World Bank literature that the large-scale and costly effort to promote zoning in Rondônia failed in practical terms due to the lack of local support. However, this situation can be reversed by creating winners in the process. For example, concentrating social and infrastructure services in high-density agricultural areas would decrease the tendency to seek control over other areas.

Overview of Conventional Analyses and Solutions

The predominant phenomenon observed in Amazonia today is a process of converting forest to pasture, in which small farmers open the way. They adopt a mining-and-mobile strategy in which they transform forest into unsustainable pasture. They do not develop other alternatives such as agroforestry because they lack the knowledge and experience, and also because the market for most products is limited. Small farmers sell the land to cattle ranchers because it is the quickest way to increase capital. The ranchers exploit this pasture for a few years and may derive some profits from cattle production. However, most of their gain is from the sale of wood on the remaining forest lands, from various fiscal incentives, and from speculation on the land itself. Increase in land values occurs close to centers of population, where infrastructure is well developed. Higher values of land encourage intensification of agriculture, which in turn takes pressure off remaining forests. However, this trend is counteracted when the government or logging companies open new roads.

In areas away from major centers of population, there is little local support for most measures, such as restriction of road-building, that would limit free access to land and natural resources. There would probably be active resistance from the local elite, who benefit most from this free access.

Proposals to counteract the deleterious effects resulting from both government incentives and from natural frontier expansion include:

- Stopping the building of new roads, at least until zoning permits a separation of areas suitable for agriculture from areas suitable for forestry or reserves
- Encouraging agroforestry through research, market support, and farmer training
- Reducing fiscal and credit incentives to ranchers
- Continuing research on land-use alternatives, especially on their economic feasibility

It is generally agreed that there is at least some possibility of carrying out these proposals. Other suggested proposals would be more difficult to implement due to strong political opposition, lack of effective enforcement of laws and regulations, and low national priority because of cost. These proposals include:

- Promoting agrarian reform both in Amazonia and in the regions of origin of most migrants (especially in northeast Brazil) to limit migration of landless people
- Collecting a tax on capital gain from land sale (in principle 25%) to discourage land speculation and establish a progressive property tax on land-holdings, with a lower rate when the land is kept in forest
- Having stronger control of logging licenses
- Developing the general level of education in the country to increase labor opportunities in industry and services
- Establishing a land-use plan (zoning), demarcating areas for intensified agriculture; sustainable wood extraction; extractivism; and protected areas for conservation, scientific research, and tourism
- Supporting small farmers by ensuring rapid titling of land, promoting specific credit and training/extension services, and establishing good social services in areas reserved for small colonists

We believe that the last two measures should be avoided, because they would only attract more migrants to Amazonia.

The LAET Diagnosis

LAET's diagnosis has been a continuous process. A first diagnosis, based on the available literature, was formulated when the program began in 1992. This diagnosis was revised after the first series of field inquiries in 1993. These inquiries were carried out in a relatively classic mode, with preelaborated questionnaires (Castellanet et al. 1998). As interactions developed between LAET and MPST (Movimento Pela Sobrevivencia da Transamazônica [Altamira]), there were regular refinements and reformulations of this diagnosis. New evidence that accumulated during action research activities in the field caused further revisions. These elements were incorporated into the documents and publications produced by the researchers of LAET (see LAET's bibliography in Appendix 2).

Initial Diagnosis

Access to Capital

The initial LAET diagnosis was based on an early survey that characterized the diversity of farmers in the region. Six types of small and medium farmers were found in the region, whose holdings ranged from 100 to 400 hectares, with an estimated yearly income of $1200 to $7990 per year (Table 9.1). The processes by which changes (trajectories) to these farming systems occurred during the last 20 years were also analyzed (Figure 9.1). The conclusion was that most farmers started with a low-income/low-capital system based on slash-and-burn and rice cultivation, then progressively accumulated capital in the form of permanent crops (mostly cacao and black pepper) or cattle. Cacao was more common in farms located on the most fertile soils, terra roxa or terra vermelha, whereas other farmers had to depend more on black pepper or coffee. Some farmers who invested in black pepper were ruined when, in 1991, the prices fell and the *Fusarium* wilt destroyed their plantations. Cacao producers were also affected by price drops and witchbroom disease during the same period, but to a lesser extent. After 1992, both cacao and livestock producers with more than 50 head of cattle obtained a reasonable income and apparently reinvested part of this income into buying more land.

The initial diagnosis was based on the assumption that economic factors were the most important determinants of farmers' choices in farming systems and that maximizing family labor return is the main strategy of family farmers. This was confirmed by the fact that the labor productivity (return per day of work) from permanent crops (cacao, pepper) and cattle were in the same range and well above rice productivity (Table 9.2).

"(text continues on page 171)"

TABLE 9.1. Characteristics (average) of the farmers identified in the 1994 survey

Type	Years on the Land	Land Area (ha)	Distance from the Road (km)	Adult Family Workers	Total Income in $US/yr	% Income*				No. of Cattle	Pasture Area (ha)
						Cacao	Pepper	Cattle	Rice		
1. Just arrived	4.0	152.5	32	4	2015	0.0	12.5	0.35	20.1	0.5	7.0
2. Pepper producer	13.5	268	14.5	3	3138	7.0	24.5	10.2	7.7	9.0	23.2
3. Cacao producer	15.0	238.5	18.5	4	7990	58.3	7.5	5.0	6.0	14.0	36.6
4. Losing capital	13.0	125	18	3	1219	0.0	3.5	3.0	22.6	1.5	5.2
5. Cacao+cattle producer	14.5	405	5.5	3.5	6962	31.0	13.0	28.0	7.4	61.0	38.1
6. Cattle producer	16.5	183	6	5.5	7562	8.0	6.0	39.3	4.4	96.5	67.5
7. Glebista	11.0	565	30	2	1489	0.9	0.1	60.0	2.1	300.0	196.6

ha = hectares; glebista = rancher who owns a medium-sized ranch.
From LAET–Diagnostico Preliminar da Agricultura Familiar na Transamazônica March 1995.
*Other sources of income include small animals, extractivism, commerce, small business and labor sale (wages).

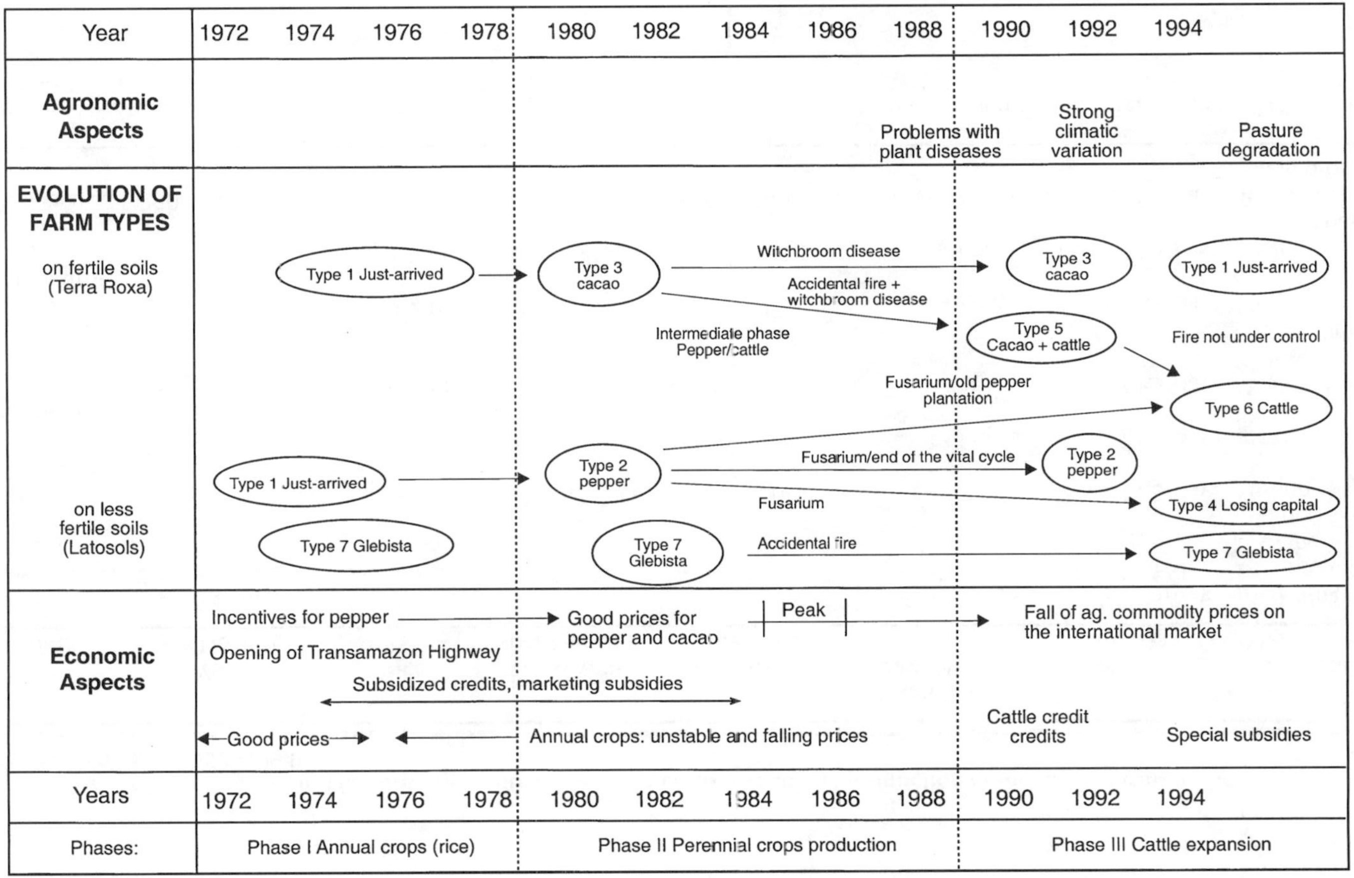

Year | 1972 | 1974 | 1976 | 1978 | 1980 | 1982 | 1984 | 1986 | 1988 | 1990 | 1992 | 1994

Agronomic Aspects

Problems with plant diseases
Strong climatic variation
Pasture degradation

EVOLUTION OF FARM TYPES

on fertile soils (Terra Roxa)

Type 1 Just-arrived
Type 3 cacao
Witchbroom disease
Accidental fire + witchbroom disease
Intermediate phase Pepper/cattle
Type 3 cacao
Type 5 Cacao + cattle
Type 1 Just-arrived
Fire not under control
Fusarium/old pepper plantation
Type 6 Cattle

on less fertile soils (Latosols)

Type 1 Just-arrived
Type 7 Glebista
Type 2 pepper
Fusarium/end of the vital cycle
Type 2 pepper
Fusarium
Type 4 Losing capital
Type 7 Glebista
Accidental fire
Type 7 Glebista

Economic Aspects

Incentives for pepper
Opening of Transamazon Highway
Good prices for pepper and cacao
Peak
Fall of ag. commodity prices on the international market
Subsidized credits, marketing subsidies
Good prices
Annual crops: unstable and falling prices
Cattle credit credits
Special subsidies

Years | 1972 | 1974 | 1976 | 1978 | 1980 | 1982 | 1984 | 1986 | 1988 | 1990 | 1992 | 1994

Phases: Phase I Annual crops (rice) | Phase II Perennial crops production | Phase III Cattle expansion

FIGURE 9.1. Evolution of farm types along the Transamazonian Highway between Altamira and Uruará. ag = agricultural; *glebista* = rancher with medium-sized ranch.

TABLE 9.2. Estimated yields and labor productivity of the main crops and animal production in the family farms in the Transamazônica (1993–1994)

Product	Yield Kg/Hectare	Price (US$)	Gross Value per Hectare	Amortization per Year	Number of Days Worked per ha	Net value of the Work Day	Observations (Main Problems)
Rice	1800 (30 bags)	4.5–9.0/bag	R$135–270	–	40–90 days	1.5–6.7 R$/day	Marketing (prices)
Cassava flour	5400 (90 bags)	5.0–8.0/bag	R$450–540	R$25.00	110–190 days	2.1–4.4 R$/day	Low productivity of the processing
Pepper (medium)	1300	0.4–0.66/kg	R$520–858	R$120 (600/5)	55 days	72–13.4 R$/day	Fusariose
Pepper (good)	1300	0.6/kg	R$1860	R$120	133 days	13.1 R$/day	Idem
Cacao (medium)	320	0.52–0.70/kg	R$166–224	R$40 (400/10)	15–20 days	6.3 R$/day	Witchbroom disease, price × quality
Cacao (good)	650	0.70/kg	R$455	Idem	45 days	9.2 R$/day	Idem
Cattle	20–50 (meat)	1.5/kg	R$30–5	R$10.00 (80/10)	5–9 days	4.0–7.2 R$/day	Degradation of the pastures
Cupuaçu	2500–5000	0.30/kg	R$500–750	R$40.00 (400/10)	45–120 days	6.2–33.3 R$/day	Transportation and processing

From DAZ/LAET (Graduate course of NEAF [European Commission]/Laboratorio Agro Ecologico da Transmazônica) field research (1994).

Diversity in the farming systems was seen as a result of initial capital or access to credit available to respective farmers, types of soils on the property, length of time since settlement, and available family labor.

Later evidence challenged the initial diagnosis in two respects: (1) other economic factors besides access to capital were important and (2) other factors besides economic ones were important when it came to deciding whether to abandon or sell the farmers' land and migrate to buy other land.

Revised Diagnosis

Other Economic Factors

Normally, farmers were reluctant to record their monetary affairs and even more reluctant to communicate this to outsiders. However, once a close network of cooperation and social relations with farmers' groups and individuals was established, in-depth observations based on farm budgets were possible. Twenty-three farmers agreed to keep an account book in which the farmer would write down all expenses and profits for the farm, whether in cash or kind. They would also record the use of the family's labor day by day, as well as the use of any hired workers. These observations were verified and analyzed by an agro-economist, who visited each farmer every month. The accounts were precious in that they permitted—perhaps for the first time in the region—first-hand information on farmers' management of time and money. They also provided an idea of the economic return from the main crops or husbandry to land, labor, and capital.

These observations demonstrated the following:

1. The variations in yield and return from one farmer to another were extremely high, ranging from one to three for most agricultural productions, including cattle.
2. The return to labor from beef cattle production was higher than from other perennial crops until 1996, but was reduced later as a result of a fall in beef prices. As commodity prices went up during the same period, it became more profitable to invest in perennial crops. The investment trend of the farmers (toward extensive beef or toward perennial crops) depended on the price ratio between these products. It was estimated that when the price of cacao locally exceeds $0.70 per kilogram, it is more economical to expand cacao than annual crops and beef cattle in order to maximize the farm's income (at an average price of $0.20/kg of paddy rice and $1.50/kg of beef (carcass) (Sablayrolles 1995).

Noneconomic Factors

Education Some farmers have a migratory strategy based on economic mining (Lena 1986), whereas others maintain a long-term presence in the region and prepare their children's future locally. It is striking that a significant proportion of the farmers today live not on farms, but in villages close to their farms. This new residence pattern seems to correlate with their desire to have their children complete primary education (Regina et al. 1995).

The farmers themselves estimated that, in one particular community, approximately half of the current landowners will not stay in the travessão (feeder road community) and are not interested in any group discussion regarding the future of the community. Medicilandia was different, however, in that it was the location of the first rural family school in the region. These schools train farmers' sons and daughters to become future farmers themselves through a system of alternate attendance. In this system, the children spend 15 days in the school located in a rural surrounding, then go back to their parents' home for another 15 days, where they help on the farm, practice what they learned, and analyze their parents' farming methods.

More than 60 families in Medicilandia asked to enlist their children in future enrollments. However, this eagerness can be interpreted in another way. For many farmers, this was the only chance to give their children a formal education, regardless of the type of education, and the real objective was to help the children find another job in town. If this is true, the rural school might, in fact, encourage rural exodus and not alleviate it.

Ecological Factors It was clear from the outset that it would be difficult to obtain unbiased views on the environment from the farmers, in either direct interviews or even in informal talks. Farmers generally claim that they intend to preserve half of their lot as forest, as required by law, although in many cases they have already deforested more. When they discussed the "farm lot of their dreams" in public meetings, they usually talked about leaving 50% of the land in forest (in some cases enriched forests) and planting the rest in diversified cropping, including agroforestry and cattle. In some cases, a few militant individuals flatly declared, "I don't give a damn about ecology and would clear out forest and expand my pastures as much as I can."

These two attitudes can be understood as the result of a conflict between immediate economic interests and a certain moral pressure from the media and government against deforestation. Researchers in general, and northern foreigners in particular, are seen as the representatives of this public reprobation, which explains the usual bias in any discussion of the environment. However, after one year of work and confidence-building by LAET, farmers were willing to discuss their true land use strategies in the presence of the researchers. Through these informal group discussions of farmers, LAET members finally cleared their doubts about the farmers' attitude toward ecology. What came from this discussion were the diverse strategies of farmers toward their land.

Some farmers declared that they wanted to keep some forest in their land for their children's sake. This confirmed two hypotheses from the research: (1) forest is the best source of fertility for future perennial crops and (2) forest also has an important potential economic value, since the value of wood is increasing regularly in the region. Farmers are interested in protecting the forest if they intend to stay and establish their children as farmers and if new lands become more difficult to acquire. Other farmers, however, preferred to convert all land into pastures, reinforcing Mahar's (1990) view on the importance of short-term "economic rationality."

Various farmers have started to plant timber species, especially mahogany (*Swietenia macrophylla*). Many were impressed by the growth of these trees in the plot of a particularly dedicated farmer who has planted 600 mahogany trees (mognos, in local terms) in his cacao plot since 1978. Today, the trees have reached 45 cm in diameter, with an estimated value to the farmer of more than $30,000. The fact is that, since 1993, loggers have started to encourage replanting of timber trees by providing free seedlings to farmers. A small but significant number of farmers have started to plant mahogany in black pepper or cacao plantations. The Altamira farmers' union was able to obtain seeds for them through an agreement with the Arara Indians, on whose reservation remained the only surviving individual mahogany seeds. Although attacks of *Hypsilla grandis* (stem borer) were common on the mahogany, this disease can be partly overcome by trimming all but one of the multiple top branches caused by the infection.

The Changing Situation in Transamazônica

LAET finally concluded that most conventional generalizations about the behavior of farmers are simplistic or no longer relevant, partly because the frontier is closing. It is closing not because land is becoming scarce, but because feeder roads cannot be maintained beyond a relatively short distance from the main road.

The mining model is correct in general terms when land is still abundant and easily accessible, and the price of beef is high in relation to cash crops. However, this is no longer the case in the Transamazônica region. As a result, many farmers have different attitudes and behavior toward natural resources. One of the main factors that explains the difference is probably the farmers' long-term perspective, especially regarding their children's future. Other factors, such as technical knowledge and management capacity, also play a role; as these improve, there is a tendency toward intensified management.

Road Accessibility

The average feeder roads now extend 38 km from the main road, although the part built by the government seldom exceeds 20 km. It is doubtful that government revenues based on agricultural commodity taxes can compensate for the cost of maintaining feeder roads beyond this point. New access roads or elongation of current roads are now opened by loggers. It is not rare for sawmills to open roads of more than 100 km to gain access to a river from which it can ship the logs, either in floating rafts or on barges, at a much cheaper cost than by road. There were various examples of these undertakings in Altamira and Uruará, where one of these roads cut through the proposed Arara Indian Reserve to the Iriri River. New lands made available by these roads were quickly occupied by small farmers, ranchers, and local businessmen. However, most of these current occupations are by farmers already settled in the region, who want to expand their land and not from newcomers. The influx of landless farmers to the region as a whole is apparently decreasing, and occupation of this land is more symbolic than real. Very few farmers live more than 30 km from the main road and new land claims, in most cases, are more speculative investments than productive ones. The land value at these distances corresponds more or less to the value of the timber wood that can be sold to loggers (e.g., $5 to $10/hectare).

Large ranchers are not interested in opening new pastures in distant areas at the moment either, probably because the price of beef has gone down in recent years and does not compensate for new investments. Ranchers have also practically stopped buying new land from small farmers in older colonized areas. Thus, the trend toward concentration of land holdings in a few hands has slowed down. After a reduction in land values in the late 1980s, prices are now stable and range from $50 to $300 per hectare in these areas, depending on the location, soil fertility, pastures, and tree plantations that have been established.

In forest lands up to 100 km from the road or from main rivers, fierce competition exists between big logging companies and small loggers to demarcate the land; that is, to mark the boundaries for a later claim. Since the 1990s, timber production has grown steadily and, today, is probably the most important source of income in the region as a whole. The big sawmill owners have accumulated considerable wealth and power during this period.

Close to rural cities or the main road, there has been an increase of very small holdings (from 3 to 10 hectares), locally called chacaras. These holdings may be located on sites initially planned for agrovilas (rural towns) by INCRA or may result from the subdivision of 100-hectare lots by original owners. These chacaras are usually situated so that access to social services and markets is much better than for more distant farmers. The small lots are often acquired by previously landless workers as a way to obtain property at an affordable cost without going to remote and isolated areas. In a survey near Uruará, 11 out of 20 chacareiros were originally landless agricultural workers. Most of them continued working outside while beginning to establish their own plantings. It is remarkable that seven of these chacareiros previously had a larger farm lot and abandoned or sold it as a result of the difficulties associated with distant access. Two other farmers tried to keep the distant farm but moved their house and family to the chacara.

For small or landless farmers, a 35- to 40-km distance from the main road seems to be the limit beyond which they will not occupy new lands, that is, if there is no regular transport. Many farmers who occupied more distant areas in the 1980s abandoned them after a few years, when they found that the road would never be improved or properly maintained.

Social Sustainability

Initially, LAET's agro-ecological research concentrated on economic indicators, mainly the annual income of the family. Income is usually considered by those using a farming-systems approach as the main factor explaining success or failure of the farm (Shaner et al. 1982). For example, in the first questionnaire given to farmers, educational problems were not treated, despite the sociologist's insistence on including them. The other team members, mostly agronomists, did not see how this question would fit into the study of agricultural problems. Through discussion with MPST, LAET came to realize the importance of these factors in the farmers' success with or abandonment of the farm. The interest and concern of the farmers for their children's education and future, as manifested during the first Altamira conference, were also strong clues for researchers.

As the team included more and more questions about education and health in its questionnaires and "diagnosis," it became clearer that education and social life were not only strong concerns of the farmers in addition to their income, but were also strong determinants in the decision to stay or leave their land holding, regardless of their income.

The influence of family on these decisions was strong. In some cases, when the children reached an age at which they had to be sent to the village to continue their education (after fourth grade), their mother decided to follow and eventually provoked the whole family to move to a more centrally located house. Many farmers presently living in a chacara are in this category. They go back to their original plot of land a few days per week if it is not too far; otherwise, they simply abandon it. In most cases, the plot becomes converted to extensive beef production, a less labor-demanding activity.

Another factor influencing farmers' decisions is social life. Many farmers frequently spend two or three days per week in town, officially to do shopping and make business contacts, but in fact simply enjoying the social life. Social life in the countryside is limited, partly because of cultural differences. Migrants are of various origins and types and have few common interests. In addition, the distance factor renders any meeting to be a time- and effort-consuming enterprise. Very often, people walk half a day to reach a meeting site, sleep there, and walk back. It became clear to LAET that the low population density (an average of one family per km^2) was a key factor in making social life difficult in this region. Without a car or regular transport, except for the truck that usually passes once a week, farmers commonly live for days and sometimes weeks without any outside contacts during the dry season. Horses or mules are used by some affluent farmers for short-distance travels, but they cannot make long trips because the animal will suffer from rapid exhaustion in the hot climate. A bicycle is a solution for young people during the dry season, but is not commonly used on the main Transamazônian road, probably because of the danger of truck drivers' irresponsible driving behavior. A motorcycle or small truck improves a farmer's life, but requires a much higher income and doesn't solve the children's schooling problem.

In the 1970s, farmers were grouped in small hamlets (agrovilas), but they quickly decided to move and build their house in their plot. Twenty years later, in the 1990s, the trend is reversed. Older farmers suggested that the initial move to the plots was motivated by a need to secure their properties (fear of having their products and cattle stolen) and by a desire to have more privacy and control of their family. Many farmers had an optimistic view of the future and believed that things would improve. For example, they thought that the roads would be paved and electricity would eventually reach the rural areas, as had happened during the last decades in the southern states. They also thought that they should grab as much land as they could, both for security and for speculation.

Today, people are seeing things differently and no longer hope that things will improve quickly. Some even fear that things will continue to deteriorate. Many farmers sold their land in 1990–1991 after losing hope for the future of the region, seeing that even the main road was no longer maintained.

Intensification of Production

Some scientists believe that reduced areas lead to a more intensive cropping pattern, which results in a decrease in soil fertility because of reduced periods of fallow. The ultimate result, they hypothesize, is a collapse of the agricultural system, because agricultural revolutions require fertilizers and because other external inputs are too expensive (Fearnside 1990b).

On the other hand, agricultural economists and historians such as Boserup (1965) demonstrated that high densities can bring a change in agricultural practices, owing to the increased availability of manpower. After a period of gradual intensification (i.e., increasing labor input and yields per unit of land occupied), the agricultural system comes into stress due to reduced fallow periods and lowered soil fertility, which brings the need and conditions for a technological revolution. Historically, the most important of these revolutions has been the passing from shifting cultivation to animal drought agriculture and introduction of legumes into the rotation, thus permitting an increase in production and yields, and the restoration of fertility (Ruthenberg 1980). The relevant point here is that agricultural revolutions can take place without artificial fertilizers or external input, a practice many ecologists criticize as inappropriate for the Amazon.

Implications of Intensification Factors relating to an increased density of farmers in Transamazônica and the intensification of land use led LAET to believe that a smaller land unit and a higher population density might actually improve the sustainability of agriculture in Transamazônica. To test this idea, LAET carried out a comparative study of 14 farms of 100 hectares, located between 20 and 50 km from the city on bad roads, and 20 chacaras of 7 hectares on average (between 3 and 20 hectares), located close (5–10 km) to the small town of Uruará. The results were surprising in that the average estimated agricultural production and the annual income were practically the same in both types of farms, despite the enormous difference in plot size. The annual average income of chacaras was 3371 reais, compared to 2861 reais for larger parcels. The difference was mostly due to the importance of perennial crops (principally black pepper) close to town, whereas in the 100-hectare plots, cattle predominated along with some grain. Chacareiros also had more opportunities to work outside their farm, either for daily wages or as meieiros (sharecroppers). Family sizes and compositions are similar in both situations (an average of 2.5 workers per family).

There may be greater availability of family labor in the chacaras, because less time is wasted going to and from the city and the family's general morale seems higher. As for the social life, all indices showed much better life and access to health and education close to town. Practically all chacareiros could send their children to secondary schools and had easy access to the city's social services, especially health care. For most, daily public transport was available. The marketing of agricultural produce also was much easier. In most distant plots, on the contrary, public transport was available less than twice a week and at a high cost, compared with farmers' incomes and the value of their agricultural products.

Soil fertility was a concern for small holders, although it had not reached the stage at which it was their main concern. Many complained of reduced yields from annual crops (although it was difficult to evaluate whether the main factor was soil fertility, increased weed pressure, or simply the lack of forest or secondary fallow for slash and burn). A few farmers initiated the use of organic manure, although still on a small scale and mostly on pepper plants. This, however, was already a revolution compared with the total absence of any form of fertilization in the 100-hectare parcels. The yields obtained in the chacaras, despite their small size, still were better or equal to the yield of more distant plots. Visual observations did not report any signs of clear fertility exhaustion or land degradation, although the soils were classified as being a "medium to low fertility class" (Ultisoils, a type of soil relatively low in fertility).

Farmers with larger plots practically never show concern for land fertility per se. They use cleared primary forest or secondary regrowth for rice production. There is little difference in average yield between the two. The technical practices and agronomic itinerary are slightly different, yet the overall average result is similar. Preparation of primary forest requires more time and a chainsaw for clearing, whereas secondary forest can be cut manually. The primary forest does not dry easily and, in some years, it cannot be burned properly before the heavy rains. In contrast, after the secondary fallow, more weeding is needed (De Reynal et al. 1995).

On the intensified plots, farmers' worries are linked much more to the effects of landscape change than to soil fertility. An increase in annual weeds in cultivated plots occurs as a result of wind dispersion of seeds from nearby plots. In pastures, the problem is bushy weeds (assa peixe, babaçu). These weeds, some of which are poisonous, can cause quick degradation from the point of view of livestock production. Noxious and poisonous weeds also cause declines in cattle productivity. Although most pastures appear to be degraded within 10 years and are often subsequently abandoned, examples of farmers who maintain reasonably productive pastures for longer than 15 years do occur.

Other research recently conducted in Marabá indicates that weeding at the appropriate time and maintaining a proper stock on the pasture are essential for pasture maintenance. Contrary to the common view, undergrazing is a factor of pasture degradation as well as overgrazing (Topall 1996). Uncontrolled fires are also a serious concern to all farmers including cattle producers, who can lose all their fodder at times when it is most needed.

Farmers in intensified plots see the disappearance of the forest as a limitation to the establishment of future perennial crops, since perennial crops grow much better after the forest than after a fallow. This is due to a higher level of soil organic matter after clearing of forest and to increased competition by weeds and bushy species in the fallowed plots. Cacao is particularly susceptible to differences between forest and fallow.

The Viewpoint of Other Stakeholders

The overall objective of the LAET project was to improve management of natural resources, with a specific goal of slowing the rate of deforestation in Transamazônica. The initial focus group comprised the farmers, since they were suspected of having the greatest impact on natural resource management. However, once researchers began to work with those in forestry and the wood industry, they realized that sawmill owners and large ranchers also strongly influenced the use of natural resources. Although these groups sometimes have interests that are in conflict with small farmers, they often have common interests. For example, sawmill owners opened roads or rehabilitated them and encouraged further occupation by landless farmers (posseiros) to cover up illegal logging and to provide cheap manpower and logistical support for loggers. For farmers who were already established, sawmill owners arranged for repair of damaged feeder roads and for provision of free rides to the city. Farmers also were able to exchange wood for the services of tractors belonging to the sawmills. Such services included the building of small dams and ponds for watering cattle.

Big ranchers also had common interests with small farmers. Ranchers offered the farmers opportunities for day labor, assistance in transport, and the renting or sharing of cattle (on loan in the half-share system). Some small farmers had a strategy of converting their land from forest to pasture and then selling it at a good profit to ranchers. Ranchers who wanted to expand their pastures quickly without having to depend heavily on contracted manpower (always seen as a "headache" by patrons) often depended on buying land from small farmers. However, ranchers in the process of expanding their ranches could be menacing to small farmers who refused to sell their land.

Merchants and service people in small towns also had an interest in the stabilization and reinforcement of family farming, since small farmers were their main customers and providers of commodities. The business community also favored small farmers, because their contribution to the population of the region was important in obtaining support from the state and federal government (through regular fiscal transfer to the counties or municípios). Such support was based on the size of the county's population.

Within urban groups, and particularly among secondary students, there is a strong condemnation of indiscriminate wood extraction and a general concern about conservation of the forest and rivers. This is probably the result of an interest in natural sites for leisure (the most popular Sunday activities are river bathing and fishing) and also of national TV programs, which often criticize indiscriminate deforestation.

Table 9.3 presents a stakeholders' analysis of various socioeconomic groups with regard to interest in conservation of natural resources and sustainable development. The analysis assumes that (1) an interest in the sustainable future of the region is closely linked with the amount of investments in buildings and land; (2) a lack of interest is influenced by a desire to relocate in a more familiar cultural situation; and (3) a short-term economic loss would result from a restricted access to natural resources.

The table indicates that other stakeholders beside farmers have an interest in sustainable development, once they give some priority to the future over immediate results. No particular group necessarily has neither short time-horizons only nor long time horizons only. However, indigenous groups, had they been included in the analyses, might have provided a contrast. But LAET had only limited contacts with one of their groups (the Arara Indians) and it was therefore difficult to generalize on this basis.

Proposals for Improving Farming Systems

LAET researchers concluded that the main problem of resource conservation in the Transamazônica region was not soil fertility, but the extension of poorly managed pastures (Schmitz et al. 1997). As a result, they suggested that a better strategy for the region would be to encourage more intensive management of smaller-sized holdings.

A comparison of social and economic factors for farmers with land holdings of different sizes suggests that 25 to 35 hectares should be sufficient

TABLE 9.3. Stakeholder analysis of various socioeconomic groups in Transamazônica

Type of Actor	Level of Local Capital	Facility to Move	Losses from Restrictions on NRM	Interest in NRM and Sustainable Development of Region
Small farmer (with perennial crops)	High	Medium	Low (some advantages also)	Medium
Small farmer (cattle only)	Medium	Medium/high	Medium	Low
Big rancher/ fazendeiro	High	High (but with losses if land prices go down)	Medium/high	Low to medium
Big sawmill owner (specialized in high-value wood)	Low (only machinery)	High	High	Very low
Local merchant or small industry	Medium/high	Medium	Medium/low	Medium
Public sector services	Low	High	Low	Medium/high (but not priority)

NRM = natural resource management

to maintain or even increase the level of agricultural production presently obtained on the average farm, based on the following cycle: 2 hectares of annual crops for one year, intercropped with leguminous cover crops, followed by 5 years of pasture, and 5 to 10 years of fallow before a new slash-and-burn cycle. This means that, at any time during a 16-year cycle on 32 hectares, there should be 10 hectares of pasture and 20 hectares of fallow. If 0.5 hectares of perennial crops were planted per year after the first annual crop, the following pattern would ensue (assuming that the perennial crop stays in production for 10 years): 2 hectares of annual crop, 7.5 hectares of pasture, 5 hectares of perennial crop, 20 hectares of fallow: total 34.5 hectares. This would guarantee a reasonable level of income based on the farm family's labor and would result in a higher population density, leading to better social services and communications.

The pattern that includes perennials is probably superior because it guarantees a reasonable income, even though livestock production is limited. However, most farmers don't have this type of rotational plan, since they give priority to increasing livestock as rapidly as possible. Instead of abandoning their pastures after five years (or when they start to become infested with weeds), they maintain cattle on them for as long as possible, resulting in severe deterioration of the productive capacity (Nepstad and Serrão 1990). Therefore, they find themselves in a "pasture crisis" in which all the forest has been cleared and all pastures are degraded, but their livestock is at its peak. They then have to sell some of their stock, and sometimes the plot itself, to buy new land (De Reynal et al. 1995).

The present chacara systems are quite recent and seem fragile, especially because of their dependence on pepper, which is subject to diseases and market fluctuation. The chacareiros have a problem of land availability for replanting new pepper plantations, since forest or high/old secondary fallow has practically disappeared from the small areas. It is likely that only through a small agricultural revolution (use of increased quantities of manure and fertilizer, animal traction or mechanization, multiple cropping, and agroforestry) might they continue their farming. Since they were poorly organized and represented in farmers' organizations, no clear request for these innovations was put before LAET.

Summary of LAET's Diagnosis

Major trends

LAET's diagnosis identified the following as major trends in the Transamazônica region:

- In some areas, there are signs of pasture expansion and the concentration of land holdings in the hands of fewer owners. At the same time, land division is occurring in older colonization areas, resulting in many mini-land holders and chacareiros who have intensified their systems of production. As a result, the total number of farm families may be still increasing.
- The social environment, especially the proximity of schools and availability of transport to small towns, has an important influence on the farmer's decision about where to live. A certain minimum population density must be achieved for farmers to have a satisfactory social life and for the State to provide minimum facilities. In most areas that have

adopted INCRA's settlement model (100 hectares per family, corresponding to 5 or 6 habitants per km^2), the human density is too low to guarantee these conditions.

- Farmers have a variety of strategies, of which the mining strategy is only one. Many farmers have one or two sons who continue as farmers on the same plot or in the same region. There is a great diversity in the level of technical knowledge and results from one farmer to the other.
- Cattle are an important component of the farmer's economy. Some farmers are in a process of cattle specialization, which may transform them into ranchers later, but this is not generally the case. Some other farmers invest in both cattle and perennial crops. The relative importance of these two types of investment depends on the relative market prices of various commodities.
- Perennial crops are important in farmers' strategies and help stabilize the family farmers' agriculture. The development of perennial crops is limited by lack of technical knowledge and managerial experience, as well as lack of initial capital or credit. The market for pepper, cacao, and coffee—the main perennial crops—is not restricted, but prices fluctuate greatly, based on world production cycles and markets.
- There are no signs of significant soil-fertility decrease at the regional level, except in specific mini-property situations. Whereas cacao is restricted to fertile soil situations, other perennials (especially black pepper) can grow well in soils considered poorer. However, the development of pastures is reaching levels at which other cultures cannot be sustained nearby, because of fire and weeds, especially in the eastern part of the region (Pacajá district).
- Diversification of crops is a good strategy for farmers to survive price fluctuations and agricultural risks, especially crop diseases.
- The local elite are not in favor of land-use restrictions, and wood extraction is now their biggest source of income. As a result, the influence of state government on resource conservation is weak. However, federal institutions are respected and can help in promoting more sustainable alternatives.

Proposals for a More Sustainable Regional Development

LAET's overall recommendation regarding the future of local agriculture in the region is to intensify and diversify the farming systems and to concentrate human population near towns, where the social and support environment is more favorable. Policies that would achieve this appear technically and economically possible, as well as socially and ecologically desirable. Proposed specific measures comprising such policies include:

- Stopping the building of new roads
- Encouraging existing perennial crops and agroforestry systems through farmer training, adequate credit, and market support
- Reducing fiscal and credit incentives to ranchers
- Modifying agrarian reform by buying and redistributing land from big estates close to the road, rather than opening new roads and lands farther from the main road
- Redistributing land based on a smaller land module (25 to 50 hectares) and providing good social services, especially schools
- Developing the general level of education in the region, especially in rural areas, and giving priority to professional training for young farmers
- Implementing participatory land-use planning (zoning) that separates areas reserved for intensified agriculture; sustainable wood extraction; extractivism; and protected areas for conservation, scientific research, and tourism
- Giving priority to regular transportation and education in the areas reserved for settlements of small colonists
- Supporting small farmers by giving them credit (contingent upon professional training) and by expediting the issue of title to their land
- Helping to provide professional training, organized by the farmers' organizations
- Consulting farmers' organizations concerning government measures and policies decided at the regional and national level

Comparison of PAR and Conventional Diagnoses

Table 9.4 summarizes and compares the conventional diagnosis of frontier dynamics with that of LAET. Table 9.5 compares conventional proposals to reduce deforestation with those of LAET. The diagnosis derived from PAR tends to be more integrated and interdisciplinary than the diagnosis derived from conventional research. The perspective of the action forces an interdisciplinary perspective on the researchers. Another difference is that the PAR proposals are more diversified and precise in their formulation than are conventional proposals. This is a result of a more-detailed and finer-grained analysis of the phenomena observed. Finally, as a result of the continuous dialogue with local people and organizations in PAR and early testing of hypotheses, unrealistic proposals tend to be discarded more quickly than those that rely on conventional diagnoses.

TABLE 9.4. Comparison of conventional diagnosis with LAET's diagnosis on the dynamics of the Amazonian Frontier and causes of deforestation

Question	Accepted Diagnosis	LAET's Diagnosis
Small farmers' strategies	Mining/mobilization	Some adopt mining strategies; others stabilize
Pasture crisis	General (except recuperation)	Quite general, but some farmers avoid it and maintain productive pastures for more than 15 years
Slash-and-burn agriculture	Ecologically unsustainable	No general fertility decline at this moment
Farmers' technical knowledge	Limited in most colonists	Extremely variable; results in highly variable economical results
Land concentration process	Rapid	Not so rapid and also accompanied by land divisions in the colonization area
Main limitations for agroforestry	Knowledge/market/ research	Market (for new tree crops), but not for existing perennials. Technical knowledge, credit
Main cause of cattle preference by small farmers	Available capital, land speculation, cultural attitude	Give good return to labor, available capital, land speculation
Cause of land sale by small farmers	Quick monetary gain, pressure from ranchers	Lack of social facilities, mostly school and transport Monetary gains in some cases
Main cause of land availability	Government roads	Tracks opened by loggers
Impact of forest industry	Growing, but less damaging to the forests than farmers	High: wood smuggling from Indian reserves, concentrated income, encourage land occupation in distant areas from the road, growing occupation of huge forest areas by logging companies

Although LAET's diagnosis confirms many of the generally accepted assumptions about the causes of deforestation in Amazonia, it differs in some important aspects; for example, in the causes of limited progress in agroforestry and of farmers' interest in cattle. Some differences may be related simply to the specific context of the Transamazonian region.

TABLE 9.5. Comparison of conventional proposals with LAET's proposals to reduce deforestation in the Frontier

Proposal	Accepted Views	LAET's Proposals
Roads	Stop building new roads	Stop building, except in agricultural area already occupied (improvement)
Agroforestry	Priority for complex systems Research Market support, training	Priority to existing perennial crops Training mostly through exchanges of experience between farmers Credit
Credit	Reduce fiscal and credit incentive to ranchers	Same, plus adapt credit for small farmers; link with training
Agrarian reform	Agrarian reform; give land title quickly to small farmers	Buy large estates close to the road and divide for landless farmers rather than organizing new settlements
Land module	?	Reduce the land module to 25–50 hectares (depending on soil type)
Land tax	Progressive land tax	Could be tried; more important: public information on land titles
Land sale taxes	Collect capital gain on sales	Not realistic
Logging	Stronger control of logging licenses	Not realistic: instead, support community and municipal forests. Authorize municipal taxes on wood products and logs.
Land use zoning	Centralized land use planning ?	Participatory land use planning involving local populations.
Support for small farmers	Credit, extension services, social services in small colonist areas (controversial)	Credit must be flexible and linked with training Encourage farmers' exchanges instead of government extension Priority for basic education in rural areas (and regular transport)
Education	Improve general level of education	Improve general level and support professional education in rural areas Alternate education (CFR)
Participation of local organizations in policies	Encourage in general	Support increased local organizations participation in municipal administration, and in regional and national public services (especially INCRA, ITERPA, IBAMA)

CFR = Casa Familiar Rural; IBAMA = Instituto Brasileiro do Meio Ambiente; INCRA = Instituto Nacional da Colonização e Reforma Agraria; ITERPA = Instituto de Terras do Pará.

However, considering that Transamazônica is often cited as an example of colonization failure and government errors in all of Brazilian Amazonia, we believe that many of these conclusions could be extended to other areas.

Identification of Applied Research Priorities

The PAR diagnosis helps to identify research priorities in applied or goal-oriented research. Such projects should help to solve thematic problems and contribute to development proposals listed earlier. They also should respond to specific farmers' demands. For example, they should deal with the following:

- Control of witchbroom disease in cacao, using low-input methods already tested by some local farmers
- Regeneration of coffee plantations after their partial abandonment
- Biological methods of fire control (living firebreaks) where crop land is adjacent to pastures
- Establishing fire-resistant timber trees in pastures and within perennial crops
- Comparing various methods of pasture rehabilitation, including the use of aggressive leguminous species (*Pueraria, Mucuna*)

Although some of these themes fall within traditional agronomic or forestry research, others may involve more fundamental studies on biological and ecological mechanisms. Pasture degradation, for example, is a complex and diverse phenomenon that needs to be investigated within an agro-ecological approach to understand the effects and outcome of rehabilitation methods. In this case, basic research should be combined with more applied research to develop appropriate technologies.

When farmers have an interest in the outcome of scientific research, they can be valuable allies of the researcher in observing nature, formulating new hypotheses, and developing ecological theory. For example, in the Venezuelan Amazon, local farmers piled up organic litter and slash around yucca plants when production began to decline, leading to the hypothesis that organic acids leached from this litter are important in mobilizing phosphorus in acidic soils (Jordan 1989).

Other examples of areas in which a deeper understanding is needed include:

- Ranchers' strategies and constraints in relation to the opening of new ranches in distant lands accessible by the river network, compared with buying land from small farmers close to roads

- Factors that determine the mobility or, on the contrary, the stabilization of migrant families in the rural sector

This discussion illustrates that, although PAR is still considered as an unconventional form of research and is looked on with suspicion by most academic organizations, it is perfectly possible to incorporate more classic academic research within a PAR program. A small PAR team can help identify interesting and exciting new subjects for thesis and research projects. Such projects can help to solve specific, sustainable development problems or, alternatively, to formulate new, challenging questions that can lead to more conventional scientific discoveries.

Evaluation of the Participatory Action Research Approach

The goal of this book has been to critique the participatory action research (PAR) method. The evaluation is based on five years' experience with the Programa Agro-Ecologico da Transamazônica (PAET) project in the Transamazonian region of Brazil. We began with a discussion of the philosophy that underlies PAR and the resource management problem to which we have applied the method (Chapters 1–4). Then, we described the project and presented the successes and failures obtained so far (Chapters 5–7). In Chapter 8, we analyzed the reasons for some of the failures and, in Chapter 9, we compared insights on the resource management problem obtained by conventional research and by PAR. In this final chapter, we evaluate the PAR approach in light of the first nine chapters and discuss the implications for developing new PAR projects.

For purposes of evaluation, we have classified the results in terms of (a) diagnosis, (b) methods of intervention, (c) process analysis, (d) links between action research and conventional research, (e) farm level results and natural resource management, and (f) scaling up to the national level.

Diagnosis

Although the process of reevaluating initial assumptions during the course of PAR is time-consuming and stressful, the resulting diagnoses provide a better understanding of the attitudes and actions of stakeholders toward management of natural resources. Consequently, proposals generated after one or more cycles of intervention and feedback have a better chance of success. Comparing the diagnosis produced by PAR with those produced by large-scale, top-down interdisciplinary studies suggests that the PAR result is better in terms of generality and realism, but certainly less precise in specific areas covered by various specialists in the projects conceived top down.

Methods of Intervention

As made evident by the experiences with the platform method and the partnership with the farmers' organization, PAR was efficient in testing and improving methods of intervention.

The Platform Method of Multiple-Stakeholder Negotiation

The platform method of multiple-stakeholder negotiation was tested in the context of municipal participatory planning. The research confirmed the potential of participatory research as a tool to facilitate a discussion by a community on its future. PAR also helps in making local stakeholders more conscious of the probable long-term consequences of present activities and practices. Through PAR, innovative proposals were formulated, which would improve natural resource management and land use in a way that would benefit the majority of the citizens. These proposals included establishing local control of fishing, creating community forests reserves, and encouraging local wood processing with low-impact technologies. Cooperation between researchers and farmers' representatives was particularly efficient when the farmers' representatives assumed the facilitation role.

The process was not successful when the government only represented the interests of a small but powerful minority. For example, in Uruará the local elite manipulated the planning process to their own advantage and against the interests of the majority of small farmers. Therefore, the multiple-stakeholder platform method was not applicable in the context of the frontier. The existence of "state of law" (passing of democratically enacted state and local laws, and their reliable enforcement) and democratic ethics is necessary for its efficiency.

In the absence of state of law, participatory research should concentrate first on reinforcing the weaker categories of the population and on analyzing political power relationships in local communities and regions.

Only later should researchers begin direct negotiations with the state and local organizations. This tactic may prove especially desirable when both the national government and local poor majorities have common interests in better natural resource management and land use. By establishing such a coalition, the capacity of the local elite to block action would be diminished.

Partnership with the Farmers' Organization

The initial assumptions on the role of the farmers' organization (MPST [Movimento Pela Sobrevivencia da Transamazônica]) in the cases described here in this cooperation were only partly confirmed:

- The farmers' organization did have an interest in sustainable development and better management of natural resources at the regional level.
- The farmers' organization effectively disseminated information in cases in which both the farmers' organization and the farmers had common interests in the proposed innovation.
- The farmers' organization played an important role in representing the farmers in other instances, such as negotiation with the State, which permitted advances in specific fields, including natural resource management.
- The farmers' organizations were an important level of collective discussion at the municipal and regional level.

However, the farmers' organization also had many other priorities and objectives, and as a result:

- The farmers' organization effectively pressured the research team. In most cases, the pressure resulted from demands of individual farmers, but in other cases it resulted from organizations whose interests were contrary to these demands.
- The farmers' organization facilitated the research in most cases, but also made research difficult or blocked it when it was contrary to its interests.

The establishment of a common strategy was not achieved. The initial model of strategy-building through the improvement of communication between farmers and researchers was found to be inappropriate. It was, therefore, not possible to conclude that the choice of the farmers' organization is the most appropriate for PAR on natural resource management in the frontier context. Farmers' organizations have both definite advantages and potential, as well as serious drawbacks and complications. Researchers cannot expect that representatives of farmers' organizations will necessarily state their own priorities clearly and expose their strategies at the beginning of the cooperation. Confidence is not easily built and needs to be gained in practice, not in rhetoric. Time is important but not necessarily the answer. Mutual distrust can also build up and render communication very difficult.

The lack of transparency and dissimulation of information on the part of farmers' organizations were observed limitations. Researchers, too, can be blamed for a lack of transparency. They explained neither their professional objectives nor their need for scientific recognition to the farmers. In the future, researchers interested in establishing partnerships with local organizations should bear in mind that, from the outset, efforts must be made to identify each organization's critical fields of interest. For example, in the cases discussed here, factors that were important to the farmers' organization but were never explained to the researchers were (a) desire for recognition by the public and by local and national institutions and (b) backing by the local farmers. Until the researchers finally understood this, they were viewed by farmers as competitors for local and national recognition. Under all circumstances, competition within the partnership organization must be avoided, whether the competition is for recognition, for funding, or for anything else.

Process Analysis

Process analysis and methodological evaluations were carried out by both the platform method and the partnership with farmers. Because there was an effort to document the processes and reflect on them critically, the LAET (Laboratorio Agro Ecologico da Transmazônica) team was able to evaluate the methods, find their weaknesses, and design improved approaches. Examples of process documentation appear in Chapter 5 and in LAET's publications, listed in Appendix 2.

Limitations of Cognitive Constructivism

Both methods tested by LAET—the establishment of a common strategy with the farmers' organizations and the platform approach to natural resource management—were based on a specific model of human interaction that has been characterized as *cognitive constructivism*. This model, based on the assumption that communication is forthright and open, suffered severe limitations in the local context. These limitations can be linked in part to the fact that various important actors, including the leaders of farmers' organizations, routinely used retention and manipulation of information as part of their personal and political strategies.

Information is a source of power, and improved communication will not necessarily result from methods such as mediation, negotiation, group dynamic, and participatory research. If no deep trust or commitment to openness exists, all these methods cannot help. Some earlier defenders of these methods, including Habermas (1984), actually acknowledged this fact, but this acknowledgment always appeared as a marginal observation applicable only to exceptional cases.

The PAET experience indicates that this exception might be the rule in many situations, especially where the concepts of citizenship and common good are not yet part of the general culture. This conclusion goes beyond the discussion on evaluation of methods of intervention and demonstrates the potential of PAR to question established paradigms in social sciences.

Linking Action Research and Basic Research

When classic research institutions see how they can benefit from associating themselves with PAR, they can become more supportive of the approach and more willing to link with PAR activities. Funding agencies interested in promoting on-farm applied research are often supportive of this type of association also, since they see PAR as a good way to drive basic researchers into more client-driven research. The experience of LAET provides an example of the evolution of cooperation between basic and action researchers.

During the first three years of its operation (1993–1995), the LAET team had little interaction with researchers from other institutions, who were using more traditional methods. LAET had frequent discussions on how the commitment to action research by their group could be compatible with individuals' need to conduct basic research to advance their careers. There were also debates, and sometimes conflicts, with young researchers who came with a research program already defined before their arrival and had no flexibility in adapting their proposals to meet the needs and priorities of the team. Therefore LAET gradually established an internal by-law, which specified the duties and rights of the team members and specified how decisions would be made. Eventually, LAET prepared a statute and registered as a nongovernment organization (NGO) in 1997.

An important aspect of the statute is that all researchers joining LAET had to agree to submit their research proposal to the team and have it discussed not only on the basis of its scientific merit, but also on its integration into the priorities of the research action program. LAET would also have to use part of its time for general interest activities, which might range from internal exchanges to training of farmers and responding to pressing demands from MPST. The possibility of accepting associate researchers, who would conduct specific research that was of interest to PAET without sharing the responsibilities and duties of LAET's permanent members, was also discussed. The key aspect of this association would be preliminary negotiation on the objectives of the research to be conducted. In return, LAET would support the proposed research in terms of infrastructure, data exchange, and other aspects.

In general, it was felt that many researchers would not accept negotiated-research topics. Negotiated research was somewhat revolutionary in regard to the established rules and customs of the academic world, which are based on disciplinary hierarchies. Some researchers even refused to discuss anything about their discipline with people from LAET, whom they considered to be outsiders. LAET discovered that one way to solve this problem was to discuss possible subjects of interest with prospective master of science or doctor of philosophy students, to negotiate a proposal with them before officially submitting it to their university. (In Brazil, prospective graduate candidates often have to present their proposal before their admission.) With this approach, various investigations could be oriented toward PAET interests. As of 1998, one doctoral dissertation (Salgado 1997) and one master of science dissertation (Sakael 1995) have been concluded, and four additional master's and doctoral dissertations were initiated.

In 1996, LAET also succeeded in negotiating specific applied research activities with some EMBRAPA (Empresa Brasileira de Pesquisa Agropecuaria) specialists. This was particularly relevant in the field of black pepper production, in which the problem was clearly identified. However, the team lacked the expertise required to test and develop technologies needed to control the *fusarium* disease. A plant pathologist from EMBRAPA, based in Belém, agreed to assist in launching a disease-control program with LAET's cooperation. Another researcher was enthusiastic about testing various green manure plants. The interest of these specialists was encouraged by the fact that LAET maintained regular contact with a network of farmers; this greatly facilitated fieldwork and direct feedback from farmers.

LAET and EMBRAPA were later able to successfully present a joint research project based on research activities for funding in a competitive grant program (Prodetab) supported by World Bank. This helped to reinforce cooperation and mutual interest. However, the PAR team had to first accumulate knowledge and competence (particularly in diagnosis of the general problem, which it tried to solve) to outweigh costs and difficulties resulting from dealing with outside researchers. The PAR team used its social network, including organizations, to facilitate this outside research. The benefits were important for both sides. The PAR team gained the thematic competencies that it needed to address specific problems and obtain concrete results in these fields. The thematic researcher gained an understanding of the broader picture and succeeded in producing a greater impact on solving applied problems.

Results at the Field Level

Sustainable Agriculture

The PAR program in Transamazônica had some successes in the field of sustainable development. Concrete results were obtained at the farm level in the development of perennial crops (black pepper), use of leguminous crops as cover crop, and establishment of a young farmers' professional training program. However, the number of farmers attained so far remains small in proportion to the size of the region and the magnitude of the problems.

Management of Forest Resources

So far, concrete results in natural resource management consist of some communities' greater resistance to selling wood at a cheap price, the establishment of community forest reserves, and local fishing restrictions in Porto de Moz. The concept of land-use planning has been popularized and adopted by the farmers' organizations as a result of the PAR program. However, changes in practices by local actors still are too limited in scale to permit any conclusions about the efficiency of PAR for solving natural resource management problems. For example, it is difficult to gauge the impact of the discussion and training on diversification and promotion of perennial crops, and on the intensification of managing pastures and soil. Also, the discussions of policies such as credit and land distribution have not yet reached a resolution.

In both sustainable agriculture and natural resource management, results are promising. Institutional or technical innovations have been developed, which have been adapted to the needs and capacities of the local people. Most innovations have a potential for large-scale replication (scaling up). The next challenge is to establish a link with policymakers to achieve this scaling up. For example, in the community forest reserves, government authorities have already been involved and have given their approval to what has been accomplished so far. The next steps would be (1) passing a law or decree to make this approach official or incorporate the approach into the existing legal framework and (2) allocating funds at the state level for a training and information program on the legal steps involved in the establishment of this new type of reserve.

Mutual Learning

Learning is an important output of PAR. The approach achieved significant results in terms of adopting new methods and concepts by the local farmers' representatives. Special mention must be made of the training component of the program, especially for young farmers in the Casa Familiar Rural and for young agronomists in the DAZ (Desenvolvimento da Agricultura Familiar Amazonica [graduate course sponsored by the European Commission]) and the agriculture degree program in Altamira. These programs prepare the next generation for new approaches in rural development.

Since PAR encouraged cumulative learning on both the local actors' and the researchers' sides, training of researchers was also a significant output of PAR.

Scaling Up to the National Level

The main advantage of PAR, compared to other research or development approaches, is its ability to produce innovative methods of intervention and validated technical and organizational proposals, adapted to local context and with a potential for large-scale adoption. Many innovations (e.g., a better-adapted credit scheme and new forms of community forests) depend on political decisions at the state or national levels. In addition, even if good technical innovations tend to be disseminated spontaneously by farmers themselves, good training and technical follow-up programs are essential to accelerate the rates of adoption. These programs require important human and financial resources. However, such resources are out of the reach and scope of action research teams, which have neither the authority nor the capacity to scale these innovations up to a national level.

PAET's initial hope was that the scaling up would be conducted by the regional farmers' organization, which was accustomed to negotiating with government (in fact, it was its main function). The organization could also mobilize substantial funds for development programs. We have seen that PAET functioned only partially. The organization functioned as a filter between researchers and the rest of the society, and exercised a censorship on the results that did not fit with its political strategy (e.g., with credit). Meanwhile, the LAET team gave priority to actual work with farmers in the field and spent limited time in lobbying and presenting its results at the regional and national levels. This strategy was a mistake that caused LAET to remain isolated from the national debates conducted by NGO professionals and funding agencies, resulting in insufficient public recognition at the national level. One consequence was a difficulty in obtaining new funds, in contrast to the experience of other organizations based in capital cities, which spent a large part of their time and efforts in marketing and lobbying. Difficulty in obtaining funds hampered the financial sustainability of LAET in the medium term. But, more important in terms of sustainable development, it also meant that obtaining support from the government for innovative proposals was more difficult.

In conclusion, it is important that a certain portion of the time and efforts are allocated to the diffusion of results and participation in national-level debates at an early stage, even if the team feels it is "still too early" to present concrete results. The problem is that when solid results are finally achieved, it might be too late to enter into the public-relations exercise. Establishing a network of contacts outside the region of work is an important part of a PAR team's job, not only to guarantee its own future but also to guarantee the scaling up of its results in the field, even if this involves a marketing approach that it might not particularly like.

Conditions for Developing New Participatory Action Research Projects

Based on the PAET experience, we offer several recommendations that might be taken into consideration in setting up new PAR programs elsewhere. These suggestions are fairly general and should apply to most PAR projects of long duration.

Institutional Considerations

In institutional terms, PAR clearly does not fall into the research patterns of existing mainstream institutions. It requires too much research for environmental and development-funding agencies and calls for long-term commitments, which they are usually not prepared to make. For research institutions, PAR does too many development and training activities, with insufficient disciplinary scientific production. However, it would be possible to overcome this difficulty by combining resources from the two sources to create "hybrid" local organizations. These would link researchers from academic institutions with development/environmental agents supported by funding agencies that are sensitive to the learning approach. Technical NGOs as well as private organizations and foundations worldwide are examples of such organizations. Government institutions concerned with applied research, development, and extension can facilitate the setting-up of such interinstitutional organizations. LAET is an example of this type of new organization.

Records and Scientific Production

PAR can produce satisfactory results only if the team makes a sufficient effort to regularly formalize its diagnosis, hypothesis, theoretical framework, and methods during the various phases of intervention. Such an effort has to be made at the beginning and end of each phase of the program. In between, special efforts must be made to accurately register the processes that occur. Recording observations is an essential tool to formalize and keep in order an accurate and objective record of the processes. As a result, a PAR team necessarily produces a great quantity of "gray literature," which has to be edited before valid conclusions and publications can be produced. This is a handicap that must be kept in mind when comparing PAR scientific production with other academic research programs.

The Action Research Team

Personal Commitment

To build the necessary relationship with local people, and also to formulate a strategy and diagnosis, there must be a fair level of continuity and commitment in a PAR team. With appropriate planning, the PAR team can become permanent in a given region, with a nucleus of at least four or five researchers fulfilling a commitment of three to four years, gradually being replaced by new members. Once the diagnosis of regional problems is well established and a clear PAR strategy is determined, other researchers with a more traditional academic orientation can join the group as associate researchers. In this case, they must agree to integrate their research objectives with PAR priorities.

Composition of the Team

Some of the nucleus researchers will probably be young, since they are less likely to have family commitments and are psychologically more open to such an experience. However, young researchers lack experience, which will further delay the formalizing and publishing process. Therefore, it is important to incorporate at least one experienced researcher into each team. This person should be familiar with the scientific world and should have experience in development interventions and methods. He or she should have broad training in an integrative discipline, such as geography, human ecology, or agricultural science, to establish a bridge between social and natural sciences. Including professional development agents in the nucleus may also be worthwhile, even if publication of the results may be delayed.

If the objective is to work closely with local organizations, it is also desirable to include an experienced anthropologist or a professional with experience working with local organizations. This person, who could also be an educated farmer, should be able to analyze the political and social strategies of the local leaders quickly and to play the role of "translator," to interpret the technoscientific language of the researchers for the farmers. Such translation is especially important in the early stages of forming an action research team, when a partnership with the farmers' organizations is sought.

Conclusions on Participatory Action Research

In conclusion, we can say that participatory action research, as experienced in the Transamazônica region, was more successful than strict disciplinary approaches in gaining an understanding of the social, economic, political, and ecological factors that affect resource management. However, in actually solving management-resource problems, the project was only partially successful within the first five years. Affecting a change on a regional scale, toward a more sustainable management of natural resources, will take considerably more time and effort.

Specifically, our conclusions are:

- PAR was an efficient tool for analyzing the causes of a given environmental problem (deforestation in a frontier region of the Amazon).
- The PAR team was able to test and improve methods of intervention to begin solving this problem and to design and test both technical and organizational solutions with local people.
- The PAR project was less efficient at scaling up these solutions, to make a significant impact on natural resource management at the regional level. To improve scaling up, both long-term continuity and better linkage of the fieldwork with national debates on policies are needed.
- The experience in Transamazônica showed that PAR in the frontier context has to be considered a long-term undertaking. For example, many significant findings were made only during the fourth or fifth year of the

program. Therefore, continuity is essential. PAR involves a substantial commitment by at least some of the researchers involved.

- PAR can be combined with conventional research on specific themes. This combination is likely to improve the efficiency of both approaches in terms of problem-solving and also to produce scientific results in various disciplinary fields.
- The PAR dynamic forces interdisciplinarity. It helped to produce a diagnosis of natural resource management, which analyzed the agronomic, economic, social, and ecological aspects of the region, and integrated them more fully than is possible with most other specialized approaches or with the current literature. Furthermore, the confrontation of unexpected difficulties produced feedback that made the formulation of new strategies of action with the farmers necessary. Interdisciplinarity and feedback make more effective proposals for management and conservation of natural resources possible.

REFERENCES

Abrunã F, Vicente-Chandler R, Pearson RW. 1964. Effect of liming on yields and composition of heavily fertilized grass and on soil properties under humid tropical conditions. *Soil Science Society of America Proceedings* 28:657–661.

Allegretti MH. 1990. Extractive reserves: An alternative for reconciling development and environmental conservation in Amazonia, pp 252–262. In: Anderson AB, ed. *Alternatives to Deforestation*. Columbia University Press, New York.

Allen T, Starr T. 1982. *Hierarchy. Perspectives for Ecological Complexity*. University of Chicago Press, Chicago.

Ammour T. 1994. Organisation de paysans des zones tropicales pour une utilisation durable des ressources naturelles: l'expérience du programme Olafo, pp 754–761. In: Symposium International: Recherches-Système en Agriculture et Développement Rural, November 21–25, 1994, Centre de Coopération Internationale en Recherche Agronomique pour le Développement, CIRAD, Montpellier, France.

Anderson AB. 1990a. Deforestation in Amazonia: Dynamics, causes and alternatives, pp 3–23. In: Anderson AB, ed. *Alternatives to Deforestation*. Columbia University Press, New York.

Anderson AB. 1990b. Extraction and forest management by rural inhabitants in the Amazonian Estuary, pp 65–85. In: Anderson AB, ed. *Alternatives to Deforestation*. Columbia University Press, New York.

Ashby J. 1986. Methodology for the participation of small farmers in the design of on-farm trial. *Agricultural Administration* 22:1–19.

Avenier MJ. 1992. Recherche action et epistemologies constructivistes, modélisation systémique et organisations socio-économiques complexes: Quelques "boucles étranges" fécondes. *Revue Internationale de Systémique* 6(4):403–420.

Bacon, Sir, F. 1994. *Anonym*. Inventeurset Scientifiques-Dictionnaire de biographies, Larousse, Paris, p. 61.

Bacow L, Wheeler M. 1984. *Environmental Disputes Resolution*. Plenum Press, New York.

Bailey RC. 1996. Promoting biodiversity and empowering local peoples in Central African forests, pp 316–341. In: Sponsel NE, Headland TE, Bailey RC, eds. *Tropical Deforestation: The Human Dimension*. Columbia University Press, New York.

Balandier G. 1967. *Anthropologie Politique*. Presses Universitaires de France, Paris.

Barbier R. 1996. La recherche-action. *Collection Anthropos*. Economica, Paris.

Bass S, Dalal CB, Pretty J. 1995. Participation in strategies for sustainable development. *Environmental Planning*. Issue no. 7 (118 pp). International Institute for Environment and Development, London.

Batmanian GJ. 1994. The pilot program to conserve the Brazilian rainforests. *International Environmental Affairs* 6(1):3–13.

Bawden RJ. 1991. System thinking and practice in agriculture. *Journal of Dairy Science* 74:2362–2373.

Bebbington A. 1991. Indigenous agricultural knowledge systems, human interests and critical analysis: Reflections on farmers' organizations in Ecuador. *Agriculture and Human Values*. 8. Winter–Spring 14–24.

Bebbington AJ, Merril-Sands D, Farrington J. 1994. Farmers and community organisation in agricultural research and extension: Functions, impacts and questions, pp 699–705. In: Symposium International: Recherches-Système en Agriculture et Développement Rural, November 21–25, 1994. CIRAD, Montpellier, France.

Bechtel W, Richardson R. 1993. *Discovering Complexity*. Princeton University Press, Princeton, NJ.

Bellon S, Mondain-Monval JF, Pillot D. 1985. Recherche-Développement et Farming System Research, à la quête de l'opérationnalité, pp 467–485. In: *Systèmes de Productions Agricoles Caribéens et Alternatives de Développement*. Développement Agricole Caraïbe Université Antilles–Guyane (Pointe à Pître, France).

Berkes F. 1990. Common-property resource management and Cree Indian fisheries in subarctic Canada, pp 66–90. In: McCay BJ, Acheson JM, eds. *The Question of the Commons: The Culture and Ecology of Communal Resources*. University of Arizona Press, Tucson.

Bertrand A, Weber J. 1995. Vers une politique nationale de gestion locale des ressources à Madagascar. 5th IASCP Conference on "Reinventing the Commons." May 24–28th. Bodo, Norway.

Bloch M. 1966. *French Rural History: An Essay on its Basic Characteristics*. University of California Press, Berkeley.

Boserup E. 1965. *The Conditions of Agricultural Growth: The Economics of Agrarian Change Under Population Pressure*. Aldine Press, Chicago.

Bourdieu P. 1994. *Raisons Pratiques. Sur la Théorie de l'Action*. Seuil, Paris.

Brightman RA. 1990. Conservation and resource depletion: The case of the Boreal Forest Algonquins, pp 121–140. In: McCay BJ, Acheson JM, eds. *The Question of the Commons: The Culture and Ecology of Communal Resources*. University of Arizona Press, Tucson.

Brokaw NVL. 1985. Treefalls, regrowth and community structure in tropical forests, pp 53–69. In: Pickett STA, White J, eds. *The Ecology of Natural Disturbances and Patch Dynamics*. Academic Press, New York.

Brossier J, Vissac B, Lemoigne JL, eds. 1990. *Modélisation systèmique et système agraire: Décision et organisation*. Institut National de la Recherche Agronomique, Paris.

Bryant CGA. 1985. *Positivism in Social Theory and Research*. St. Martins Press, New York.

Bushbacher R, Uhl C, Serrão S. 1988. Abandoned pastures in eastern Amazonia: Nutrient stocks in the soil and vegetation. *Journal of Ecology* 76:663–699.

Campbell A. 1994. Land care in Australia: Spawning new models of inquiry and learning for sustainability, pp 366–370. In: Symposium International Recherches-Système en Agriculture et Développement Rural, November 21–25. CIRAD, Montpellier, France.

Carroll RW. 1992. *The Development, Protection and Management of the Dzangha-Sangha Dense Forest Special Reserve*. World Wildlife Fund (USA), Bangui, Central African Republic.

Castellanet C. 1992. Recherches sur l'environnement ou recherche-formation pour l'environnement et le développement. *Courrier de la Cellule Environnement de I'INRA* 15:61–65.

Castellanet C, Alves J, David B. 1996. A parceria entre organizações de produtores e equipe de pesquisadores. *Agricultura Familiar: Pesquisa, Formação e Desenvolvimento* 1(1).

Castellanet C, Simões A, Celestino Filho P. 1998. *Diagnostico Preliminar da Agricultura Familiar na Transamazônica: Indicações para Pesquisa e Desenvolvimento*. Documentos no 105 (48 pp). EMBRAPA/CPATU (Empresa Brasileira de Pesquisa Agropecuária/Centro de Pesquisa Agropecuária dos Trópicos Umidos). Belém (Pará) Brasil.

CEDI (Centro Ecumênico de Documentação e Informação). 1993. O "Ouro Verde" das Terras dos Índios (exploração empresarial de madeira de lei em áreas indígenas da Amazônia brasileira). Relatório não publicado, 52 pp.

Chambers R, Pacey RA, Thrupp LA, eds. 1989. *Farmers First: Farmer Innovation and Agricultural Research*. Intermediate Technology Publications, London.

Checkland P, Scholes J. 1990. *Soft System Methodology in Action*. John Wiley, Chichester, UK.

Colchester M. 1995. Nature savage, nature suave? Peuples indigènes, zones protegées et conservation de la biodiversité. UNSRID Discussion papers: UNSRID (United Nations Research Institute for Social Development), Genève, Suisse.

Collinson MC. 1983. *Farm Management in Peasant Agriculture*. Westview Press, Boulder, CO.

Collinson MC. 1988. The development of African Farming System Research: Some personal views. *Agricultural Administration* 22:7–22.

Comte A. 1854. Système de politique positive. Politique d' Auguste Comte. Textes choisis et présentés par Pierre Arnaud (1965). Armand Colin, Paris.

Conway GR. 1985. Agroecosystems analysis. *Agricultural Administration* 20:31–35.

Cornwall A. 1992. Tools for our trade? Rapid or participatory rural appraisal and anthropology. *Anthropology in Action* 13:12–14.

Coy M. 1996. Différenciation et transformation de l'espace au nord du Mato Grosso. Contribution à un modèle dynamique des fronts pionniers en Amazonie brésilienne, pp 103–129. In: Albaladejo C, Tulet JC, eds. *Les fronts pionniers en Amazonie brésilienne*. L'Harmattan. Paris.

Crozier M, Friedberg E. 1977. *L'Acteur et le Système*. Seuil, Paris.

Dantas M, Muller NRM. 1979. Aspectos fito-Sociologicos da Mata sobre terra roxa na região de Altamira. *Anais da Sociedade de Botamica do Brasil* 30:205–218.

Davey S. 1993. Creative communities: Planning and comanaging protected areas. In: Kemf E, ed. *Indigenous Peoples and Protected Areas: The Law of Mother Earth*. Earthscan Publishers, London.

De Reynal V, Muchagata M, Topall O, Hebette J. 1995. Agricultures familiales et développement en front pionnier amazonien. Laboratorio Agro Socio Ambiental do Tocantins: Groupe de Recherche et d'Echanges Technologiques–Université Antilles-Guyanne, Paris.

Duhem P. 1914. *La Théorie Physique: Son Objet, sa Structure*. M. Rivière, Paris.

Eden M. 1987. Traditional shifting cultivation and the tropical forest ecosystem. *Trends in Ecology and Evolution* 2:340–343.

Emery FE, Thorsrud E. 1976. *Democracy at Work: The Report of the Norwegian Industrial Democracy Program*. M. Nijhoff Press, Leyden, The Netherlands.

Falesi IC. 1976. Ecossistema de Pastagem Cultivada na Amazonia Brasileira. *Boletim Técnico*. CPATU/EMBRAPA Belém.

Fall AS, Lericollais A. 1992. Light, rapid rural appraisal: Des méthodologies brillantes et légères? *Bulletin de l'APAD*. 3:9–15.

Fals-Borda O, Rahman MA. 1991. *Action and Knowledge. Breaking the Monopoly with Participatory Action-Research*. Apex Press, New York.

FAO (Food and Agriculture Association). 1995. Forest resource assessment 1990—global synthesis. FAO Forestry Paper no. 124. Rome.

Fearnside PM. 1980. Os effeitos das pastagens sobre a fertilidade do solo na Amazonia brasileira. *Acta Amazonica* 10:119–132.

Fearnside PM. 1982. Alocação do uso da terra dos colonos da Rodovia Transamazonica e sua relação com a capacidade de suporte humano. *Acta Amazonica (Manaus)* 12(3):549–578.

Fearnside PM. 1986. Alternativas de desenvolvimento na amazônia brasileira: Uma avaliação economica. *Ciencia e Cultura* 38(1):37–59.

Fearnside PM. 1990a. Predominant land uses in Brasilian Amazonia, pp 231–245. In: Anderson AB, ed. *Alternatives to Deforestation*. Columbia University Press, New York.

Fearnside PM. 1990b. Estimation of human carrying capacity in rainforest areas. *TREE* 5(6):192–196.

Fearnside PM. 1991. Desmatamento e desenvolvimento agricola da Amazônia, pp 207–222. In: Lena P, Engracia de Oliveira A (org). *Amazônia. A fronteira agricola 20 anos depois.* Museu Goeldi, Belém, Brasil.

Feyerabend P. 1975. *Against Method.* Thetford, London.

Floquet A, Mongbo R. 1994. Savoirs locaux et approches systèmes: L'exemple d'innovations endogènes au Sud du Benin, pp 603–606. In: Symposium International: Recherches-Systèmes en Agriculture et Développement Rural, November 21–25, 1994. CIRAD, Montpellier, France.

Fox J. 1992. Democratic rural development: Leadership accountability in regional peasant organizations. *Development and Change* 23(2):1–36.

Freeman RE. 1984. *Strategic Management: A Stakeholder Approach.* Pitman, Boston.

Freire P. 1970. *Pedagogy of the Oppressed.* Harper & Herder, New York.

Fujisaka S. 1989. A method for farmers' participatory-research and technology transfer: Upland soil conservation in the Philippines. *Experimental Agriculture* 25:423–433.

Fujisaka S. 1991. Thirteen reasons why farmers do not adopt innovations intended to improve the sustainability of upland agriculture, pp 509–522. In: *Evaluation for Sustainable Land Management in Developing World.* ISBRAM, Bangkok.

Geffray C. 1995. Chroniques de la servitude en Amazonie brésilienne, 195 pp. Karthala, Paris.

Giddens A. 1979. *Central Problems in the Social Theory: Action, Structure and Contradiction in Social Analysis.* Macmillan, London.

Glaser BG, Strauss AL. 1967. *The Discovery of Grounded Theory: Strategies for Qualitative Research.* Aldine Press, Chicago.

Godard O. 1992. La relation interdisciplinaire: Problèmes et stratégies, pp 427–456. In: Jollivet M, ed. *Sciences de la Nature, Sciences de la Société.* Centre National de la Recherche Scientifique, Paris.

Grimble R, Wellard K. 1997. Stakeholders' methodologies in natural resources management: A review of principles, contexts, experiences and opportunities. *Agricultural Systems* 55(2):173–193.

Habermas J. 1984. Theory of communicative action, vol 1. In: *Reason and the Rationalization of Society.* Beacon Press, Boston.

Hamelin P. 1990. Occupation humaine le long de la Transamazonienne: Le cas de Uruará. *Cahiers du Brésil Contemporain* 11:77–94.

Hannah L. 1992. *African People, African Parks: An Evaluation of Development Initiatives as a Means of Improving Protected Areas Conservation in Africa.* Conservation International, Washington, DC.

Hardin G. 1968. The tragedy of the commons. *Science* 162:1243–1248.

Harsthorn GS. 1990. Natural forest management by the Yanesha Forestry Cooperative in Peruvian Amazonia, pp 128–138. In: Anderson AB, ed. *Alternatives to Deforestation.* Columbia University Press, New York.

Hebette J. 1994. O MPST, passado e futuro, glórias e desafios. A historia não se repete, 7 pp. Universidade Federal do Pará (Belém)/Centro Agroambiental do Tocantins Marabá.

Hebette J. 1996. Relações pesquisadores agricultores. Uma analise estrutural. *Agricultura Familiar* 1(1):39–57.

Hecht S. 1984. Cattle ranching in Amazonia: Political and ecological considerations. In: Schmink M, Wood CH, eds. *Frontier Expansion in Amazonia.* University of Florida Press, Gainsville.

Hecht S. 1993. The logic of livestock and deforestation in Amazonia. *Bioscience* 43(10): 687–695.

Hopkins MS, Graham AW. 1984. Viable soil seed banks in disturbed lowland rainforest sites in North Queensland. *Journal of Ecology* 9:71–79.

IBDF (Instituto Brasileiro de Desenvolvimento Florestal). 1975. *Inventario Florestal da Rodavia Transamazônica*. Ministerio da Agricultura, Belém, Brazil.

IPCC (Intergovernmental Panel on Climate Change). 1990. *Climate change: The IPCC scientific assessment*. Cambridge University Press, Cambridge, UK.

Ison RL, Maiteny PT, Carr S. 1997. Systems methodology for sustainable natural resources research and development. *Agricultural Systems* 55(2):257–272.

Jiggins J, Roling N. 1997. Action research in natural resource management. *Etudes et Recherches sur les Systèmes Agraires et le Développement* 30:151–167.

Johns AD. 1988. Effect of selective timber extraction on rain forest structure and composition and some consequences for frugivore and folivores. *Biotropica* 20:31–37.

Jordan CF. 1987. Shifting cultivation, pp 7–23. In: Jordan CF, ed. *Amazonian Rain Forests. Ecosystem Disturbance and Recovery*. Springer-Verlag, New York.

Jordan CF. 1989. An Amazonian rain forest. The structure and function of nutrient stressed ecosystem and the impact of slash and burn agriculture. Man and Biosphere/UNESCO (United Nations Educational, Scientific and Cultural Organization). Parthenon, Carnforth, UK.

Jordan CF, Miller C. 1996. Scientific uncertainty as a constraint to environmental problem solving: Large-scale ecosystems, pp 91–117. In: Lemmons J, ed. *Scientific Uncertainty and Environmental Problem Solving*. Blackwell, Cambridge, UK.

Jordan WR III, Gilpen ME, Aber JD. 1987. Restoration ecology: Ecological restoration as a technique for basic research, pp 2–21. In: Jordan WR, Gilpin ME, Aber JD, eds. *Restoration Ecology*. Cambridge University Press, Cambridge, UK.

Kemf E. 1993. *The Law of the Mother*. Sierra Club Books, San Francisco.

Korten DC. 1980. Community organization and rural development: A learning process approach. *Public Administration Review* 40:480–511.

Kuhn TS. 1977. *The Essential Tension*. University of Chicago Press, Chicago.

Kumari K. 1996. Sustainable forest management: Myth or reality? Exploring the prospects for Malaysia. *Ambio* 25(7):459–467.

Laet. 1994. 1° Seminário de planificação Estratégica do PAET. Agosto 1994. Altamira, Brazil.

Lakoff SA. 1980. Ethical responsibility and the scientific vocation, pp 19–32. In: Lakoff SA, ed. *Science and Ethical Responsibility*. Addison-Wesley Reading, MA.

Lal R. 1991. Myths and scientific realities of agroforestry as a strategy for sustainable management for soils in the tropics. *Advances in Soil Science* 15:91–137.

Latour B, Woolgar S. 1986. *Laboratory Life: The Construction of Scientific Facts*. Princeton University Press, Princeton, NJ.

Ledec G, Goodland R. 1989. Epilogue: Perspective on tropical land settlement. In: Partridge WL, Schumann DA, eds. *The Human Ecology of Tropical Land Settlement in Latin America*. Westview Press, Boulder, CO.

Legay JM, Deffontaines JP. 1992. Complexité, observation et expérience. In: Jolivet M, ed. *Sciences de la Nature, Sciences de la Société*. Centre National de la Recherche Scientifique, Paris.

Leite LL, Furley PA. 1985. Land development in the Brazilian Amazon with particular reference to Rondônia and the Ouro Preto colonisation project, pp 119–139. In: Hemming JI, ed. *Change in the Amazon Basin*, vol 2. *The Frontier after a Decade of Colonization*. Manchester University Press, Manchester, UK.

Le Moigne JL. 1984. *La Théorie du Système Général, Théorie de la Modélisation*. Presses Universitaires de France, Paris.

Lena P. 1986. Aspects de la frontière amazonienne. In: Frontières, mythes et pratiques (Brésil, Nicaragua, Malaysia). *Cahiers de Sciences Humaines* 22(3–4):319–343.

Leopold A. 1949. *A Sand County Almanac, and Sketches Here and There*. Oxford University Press, New York.

Levins R, Lewontin R. 1980. Dialectics and reductionism in ecology, pp 107–137. In: Saarinen E, ed. *Conceptual Issues in Ecology*. Reidel, Dordrecht, The Netherlands.

Levins R, Lewontin R. 1985. *The Dialectical Biologist*. Harvard University Press, Cambridge, MA.

Lewin K. 1946. Action research and minority problems. *Journal of Social Issues* 2:34–46.

Liu M. 1990. Problèmes posés par l'administration de la preuve dans les sciences de l'homme. *Revue Internationale de Systémique* 4(2):267–294.

Liu M. 1992. Présentation de la recherche-action: Definition, déroulement et resultants. *Revue Internationale de Systémique* 6(4):293–311.

Liu M. 1997. *Fondements et Pratiques de la Recherche-Action*, 351 pp. L'Harmattan, Paris.

Long N, Long A. 1992. *Battlefields of Knowledge. The Interlocking of Theory and Practice in Social Research and Development*. Routledge, London.

Mahar DJ. 1990. As politicas governamentais e o fesmatamento na tegiao Amazônica do Brasil. In: Bologna G, ed. *Amazônia, Adeus*, pp 69–131. Nova Fronteira, Rio de Janeiro.

Malinowski B. 1949. *Toward a Codification of Functional Analysis in Social Theory and Social Structure*. The Free Press, New York.

Maser C. 1996. *Resolving Environmental Conflict. Toward Sustainable Community Development*. St Lucie Press, Delray Beach, FL.

Mazoyer M. 1986. Rapport de synthèse préliminaire du comité dynamique des systèmes agraires. Ministère de la recherche, Paris.

McC. Netting R. 1993. *Smallholders, Householders: Farm Families and the Ecology of Intensive, Sustainable Agriculture*. Stanford University Press, Stanford, CA.

McCay BJ, Acheson JM. 1990. Human ecology of the commons, pp 1–34. In: McCay BJ, Acheson JM, eds. *The Question of the Commons: The Culture and Ecology of Communal Resources*. University of Arizona Press, Tucson.

Mc Kinnon J, Mc Kinnon K, Child G, Thorsell J. 1990. *Aménagement et Gestion des Aires Protégées Tropicales*. IUCN International Union for the Conservation of Nature, Gland, Suisse.

Meadows DH, Meadows DL, Randers J, Behrens WW III. 1972. *Halte à la Croissance* (The Limits of Growth). Fayard, Paris.

Medeiros P, Federicci A, Souza P. 1995, Abril. Reflexões acerca de uma pesquisa participativa. Communicação ao Seminario NEAF "relação agricultores-pesquisadores," 3 pp. Belém, Brazil.

Meillassoux C. 1975. *Femmes, Greniers et Capitaux*. Maspero, Paris.

Merrill Sands D, Collion MH. 1993. Making the farmers' voice count: Issues and opportunities for promoting farmer-responsive research. *Journal for Farming System Research-Extension* 4(1):139–161.

Mitchell JC. 1983. Case and situation analysis. *Sociological Review* 31(2):187–211.

Mongbo RL, Floquet A. 1994. Systèmes de connaissances agricoles et organisations paysannes au Benin: Les limites des approches systémiques, pp 744–747. In: Symposium International: Recherches-Système en Agriculture et Développement Rural, November 21–25. CIRAD, Montpellier, France.

Monteiro R. 1996. Informação e redes de interação no novo ciclo de mobilização dos pequenos agricultores da Transamazônica. Diss. de Mestrado PLADES. Universidade Federal do Pará/Nucleo de Altos Estudos Amazônicos, Belém, Brazil.

Moran EF. 1981. *Developing the Amazon*. Indiana University Press, Bloomington.

Moran EF. 1989. Government-directed settlement in the 1970's: An assessment of Transamazonian Highway colonization. In: Partridge WL, Schumann DA, eds. *The Human Ecology of Tropical Land Settlement in Latin America*. Westview Press, Boulder, CO.

Moran EF. 1996. Deforestation in the Brasilian Amazon, pp 149–164. In: Sponsel NE,

Headland TE, Bailey RC, eds. *Tropical Deforestation—The Human Dimension*. Columbia University Press, New York.

Morin E. 1991. *La Méthode*, 4 vols. Seuil, Paris.

Moser W, Peterson J. 1981. Limits to Obergurgl growth. *Ambio* 10(2–3):68–72.

MPST (Movimento Pela Sobrevivencia da Transamazônica). 1991. Projeto global de desenvolvimento da região Transamazônica. Trecho Repartimento-Ruropolis, 32 pp. MPST, Altamira-Pará, Brazil.

MPST (Movimento Pela Sobrevivencia da Transamazônica). 1993. Breve historico do Movimento Pela sobrevivencia da Transamazônica, 7 pp. MPST, Altamira-Pará, Brazil.

Mt Pleasant J. 1990. Weed population dynamics and weed control in Peruvian Amazon. *Agronomy Journal* 82:102–112.

Myers N. 1979. *The Sinking Arch*. Pergamon Press, Oxford, UK.

Myers N. 1984. *The Primary Source: Tropical Forests and Our Future*. Norton, New York.

Nelson N, Wright S. 1995. Participation and power, pp 1–18. In: Wright S, Nelson N, eds. *Power and Participatory Development: Theory and Practice*. Intermediate Technical Publishers, London.

Nepstad D, Uhl C, Serrão EA. 1990. Surmounting barriers to forest regeneration in abandoned, highly degraded pastures: A case study from Paragominas, pp 215–229. In: Anderson AB, ed. *Alternatives to Deforestation*. Columbia University Press, New York.

Nicholaides JJ, Bandy DE, Sanchez PA, Valverde CS. 1982. Continuous cropping potential in the Amazon, pp 337–365. In: Schmink M, Wood C, eds. *Frontier Expansion in Amazonia. Center for Latin American Studies*. University of Florida, Gainesville.

Nye PH, Greenland DJ. 1960. The soil under shifting cultivation. Technical Communications, no. 51, Commonwealth Agricultural Bureaux, Farnham Royal, Bucks, UK.

O'Brien WE, Flora CB. 1992. Selling appropriate development vs. selling out rural communities: Empowerment and control in indigenous knowledge discourse. *Agriculture and Human Values Spring*, 95–102.

Olivier De Sardan J-P. 1995. *Anthropologie et Développement*. Karthala, Paris.

Olivier De Sardan J-P. 1996. Les approches participatives en matière de développement rural. Point de vue des sciences sociales. Conference du 26 November (12 pp) CNEARC (Centre National d'Etudes Agronomiques des Régions Chaudes), Montpellier, France.

Ollagnon H. 1989. Une approche patrimoniale du milieu naturel. In: Mathieu N, Jollivet M, eds. *Du Rural à l'Environment*. Institut National de la Recherche Agronomique. L'Harmattan, Paris.

Orr DW. 1992. *Ecological Literacy. Education and the Transition to a Postmodern World*. State University of New York Press, Albany, New York.

Ostrom E. 1990. *Governing the Commons: The Evolution of Institutions for Collective Action*. Cambridge University Press, Cambridge and New York.

Ostrom E, Gardner R, Walker J. 1994. *Rules, Games and Common-Pool Resources*. University of Michigan Press, Ann Arbor.

Ozorio de Almeida AL. 1992. *Colonização Dirigida na Amazonia*. Ipea, Rio de Janeiro.

Parsons T. 1951. *The Social System*. The Free Press, New York.

Piaget J. 1972. *Problèmes de Psychologie Génétique*, 174 pp. Denoël, Paris.

Pimbert M, Gujja B, Shah M. 1996. Village voices challenging wetland management policies: PRA experiences from Pakistan and India, pp 37–41. In: PLA Notes, no. 27. International Institute for Environment and Development, London.

Pivot A, Perocheau A. 1994. Le fonctionnement des programmes de recherche interdisciplinaires. Natures Sciences Sociétés–Dialogues, Bulletin no. 6:1–4, Paris.

Pomeroy LR, Hargrove EC, Alberts JJ. 1988. The ecosystem perspective. In: Pomeroy LR, Alberts JJ, eds. *Concepts of Ecosystem Ecology*. Springer-Verlag, New York.

Popper KR. 1983. *Realism and the Aim of Science*. Rowman and Littlefield, Totowa, NJ.

Rabelais F. 1964. *Pantagruel*. Chapter 8, 137 pp. First edited by François Juste in Lyon (France) in 1542. New Edition: Gallimard. Paris, France.

Rahman MA. 1993. *People's Self Development. Perspectives on Participatory Action Research*. Zed Books, London.

Ravetz J. 1989. *The Merger of Knowledge with Power. Essays in Critical Science*. Mansell, London.

Redford KH, Maclean Stearman A. 1993. Forest dwelling native Amazonians and conservation. *Conservation Biology* 7:248–255.

Regina M, Rocha CG, Zaquieu JH, Albuquerque J. 1995. *Estudo da Dinâmica de Funcionamento da Localidade 110 N, Medicilândia*. Universidade Federal do Pará. (Belém)/Nucleo de Estudo da Agricultura Familiar (Belém)/Belém, Pará, Brasil.

Rhoades RE. 1984. *Breaking New Ground: Agricultural Anthropology*. International Potato Center, Lima, Peru.

Rhoades RE. 1986. Using anthropology in improving food production. Problems and prospects. *Agricultural Administration* 22:57–78.

Rhoades RE, Booth R. 1982. Farmer-back-to-farmer: A model for generating acceptable technology. *Agricultural Administration* 11:127–137.

Rocha C, Castellanet C, Mello R. 1996. Diagnóstico rápido participativo do município de Porto de Moz: Recursos naturais (polycop), 34 pp and annexes. LAET (Laboratorio Agro Ecologico da Transmazônica), Altamira, Brazil.

Roger EM, Kincaird DL. 1981. *Communication Networks: Toward a New Paradigm for Research*. The Free Press, New York.

Roling N. 1988. *Extension Science*. Cambridge University Press, Cambridge, UK.

Roling N. 1994. Creating human platforms to manage natural resources: First results from a research program, pp 391–395. In: Symposium International: Recherches-Système en Agriculture et Développement Rural, November 21–25. CIRAD, Montpellier, France.

Roling N. 1996. Toward an interactive agricultural science. *European Journal of Agricultural Education and Extension* 2(4):35–48.

Roling N, Engel P. 1992. The development of the concept of Agricultural Knowledge Information Systems (AKIS), pp 125–137. In: Rivera WM, Gustafson D, eds. *Agricultural Extension*. Elsevier, Amsterdam.

Roqueplo P. 1996. *Entre Savoir et Décision, l'Expertise Scientifique*. Institut National de la Recherche Agronomique, Paris.

Rosenberg A. 1988. *Philosophy of Social Science*. Westview Press, Boulder, CO.

Rotblat J. 1982. The movement of scientists against the arms race, pp 115–160. In: Rotblat J, ed. *Scientists: The Arms Race and Disarmament*. Taylor & Francis, London.

Ruthenberg H. 1980. *Farming Systems in the Tropics*. Clarendon Press, Oxford, UK.

Sablayrolles P. 1995. Tipologia de sistemas de produção como subsídio a definição de politicas agricolas: O caso da agricultura na Transamazônica (45 pp and annexes). Food and Agriculture Organization (FAO)/Instituto Nacional da Colonização e Reforma Agraria (INCRA). Universidade Federal do Pará, Belém, Brazil.

Sahlins M. 1989. *Des Iles dans l'Histoire*. Gallimard, Paris.

Sakael K. 1995. La culture du poivre en Amazonie brésilienne: Le cas de la micro région d'Altamira (Amazonie brésilienne). Mémoire de Diplôme d'Agronomie Approfondie. Centre National d'Etudes Agronomiques des Régions Chaudes, Montpellier, France.

Salati E. 1990. Amazonia, pp 479–493. In: Turner BL II, ed. *The Earth as Transformed by Human Action*. Cambridge University Press, New York.

Saldarriaga JG. 1988. Long term chronosequence of forest succession in the Upper Rio Negro (Venezuela). *Journal of Ecology* 76:938–958.

Salgado I. 1995. *Relatório de pesquisa sobre a exploração madeireôira feita em Uruará*. Documento interno LAET (Laboratorio Agro Ecologico da Transmazônica)(Altamira) Pará, Brazil.

Salgado I. 1997. L'exploitation et la conservation de *Cedrela odorata, Carapa guianensis* et

Swietenia macrophylla (Meliaceae) en Amazonie brésilienne. Thèse de doctorat. October 1997. University of Paris, Paris.

Salgado I., Castellanet C. 1997. Recherche participative et planification locale pour l'utilisation des ressources forestières. Le cas du municipe d'Uruará en Amazonie brésilienne. Communication to the NEAF Seminar of Marabá, Universidade Federal do Pará/Nucléo de Estudos da Agricultura Familiar. Belém, Brazil, March 1997.

Salmon WC. 1992. Scientific explanation. In: *Introduction to the Philosophy of Science*. Prentice-Hall, Englewood Cliffs, NJ.

Sanchez PA. 1976. *Properties and Management of Soils in the Tropics*. John Wiley, New York.

Sanchez PA. 1987. Management of acid soils in the humid tropics of Latin America, pp. 63–107. In: Sanchez PA, Stoner ER, and Pushparajah A. (Eds). *Management of acid tropical soils for sustainable agriculture*. International Board for Soil Research and Management.

Sawyer D. 1990. The future of deforestation in Amazonia: A socioeconomical and political analysis, pp 265–274. In: Anderson AB, ed. *Alternatives to Deforestation*. Columbia University Press, New York.

Sayer J. 1991. Rainforest buffer zones: Guidelines for protected areas managers. International Union for the Conservation of Nature, Cambridge, UK.

Schaffner KF. 1992. Philosophy of medicine. In: *Introduction to the Philosophy of Science*. Prentice-Hall, Englewood Cliffs, NJ.

Schmitz H, Simöes A, Castellanet C. 1997. Why do farmers experiment with animal traction in Amazonia?, pp 177–198. In: *Farmers Research in Practice. Lessons from the Field*. Institute for Low External Input Agriculture (Wageningen)/Intermediate Technology Publications, London.

Schneider RR. 1995. Government and the Economy on the Amazon Frontier. World Bank. Environment Paper no. 11. World Bank, Washington DC.

Scoones I, Thompson J. 1994. *Beyond Farmers First*. Intermediate Technology Publications, London.

Scott GAJ. 1987. Shifting cultivation where land is limited, pp 34–45. In: Jordan CF, ed. *Amazonian Rainforest*. Springer-Verlag, New York.

Serrão EAS, Falesi I, Da Veiga J, Teixiera Neto JF. 1978. Productivity of cultivated pastures on low fertility soils in the Amazon, pp 195–225. In: Sanchez PA, Tergas LE, eds. *Pasture Production in the Acid Soils of the Tropics*. CIAT (Centro Internacional de Agricultura Tropical) Cali, Colombia.

Shaner WW, Phillipp PF, Schmehl WR. 1982. *Farming Systems Research and Development. Guidelines for Developing Countries*. Westview Press, Boulder, CO.

Simon HA. 1973. The organization of complex systems, pp 1–28. In: Pattee H, ed. *Hierarchy Theory. The Challenge of Complex Systems*. G. Braziller, New York.

Simon HA. 1981. *The Sciences of the Artificial*. MIT Press, Cambridge, MA.

Skole D, Tucker C. 1993. Tropical deforestation and habitat fragmentation in the Amazon: Satellite data from 1978 to 1988. *Science* 260:1905–1909.

Staver C. 1991. The role of weeds in the productivity of Amazonian bush fallow agriculture. *Experimental Agriculture* 27:287–304.

Stoecker R. 1991. Evaluating and rethinking the case study. *Sociological Review* 39(1):88–112.

Subler S, Uhl C. 1990. Japanese agroforestry in Amazonia: A case study in Tome Açu, Brazil, pp 152–166. In: Anderson AB, ed. *Alternatives to Deforestation*. Columbia University Press, New York.

Susskind L, Cruikshank J. 1987. *Breaking the Impasse: Consensual Approach to Resolving Public Disputes*. Basic Books, New York.

Taylor J, Johansson L. 1997. Nos voix, nos paroles et nos images. Projets, vérités et cassettes videos venus de la zone protégée de Nogorongoro. *Bull Arbres, Forêts et Communautés Rurales*. 10; 4–15.

Thill G. 1991. The barbecho crisis: Revisited. CIAT (Centro de Investigacion Agricola Tropical) British Tropical Agriculture Mission. Technical Report no. 1. Santa Cruz, Bolivia.

Thorsrud E. 1972. Policy making as a learning process. In: Cherns AB, Sinclair R, Jenkins WI, eds. *Social Sciences and Government*. Tavistock Publications, London.

Toledo JM, Serrão EA. 1982. Pasture and animal production in Amazonia. In: *Amazonia, Agriculture and Land Use Research*. Susanna Hecht (compiladora). CIAT (Centro de Investigacion Agricola Tropical) Cali, Colombia.

Topall O. 1996. L'arbre et l'herbe en zone tropicale humide. Gestion des pâturages sur une frontière agricole amazonienne dans la région de Marabá, pp 260–265. In: Pichot J, ed. *Fertilité du Milieu et Stratégies Paysannes sous les Tropiques Humides*. CIRAD, Montpellier, France.

Trist ABE, Higgin G, Murray H, Pollock A. 1963. *Organizational Choice*. Tavistock Publications, London.

Uhl C, Buschbacker R, Serrão EAS. 1988. Abandoned pastures in Eastern Amazonia. *Journal of Ecology* 76:663–681.

Uhl C, Jordan C. 1984. Succession and nutrient dynamics following forest cutting and burning in Amazonia. *Ecology* 65(5):1476–1490.

Verissimo A, Barreto P, Mattos M, Tarifa R, Uhl C. 1992. Logging impacts and prospects for sustainable forest management in an old Amazonian frontier: The case of Paragominas. *Forest Ecology and Management* 55:169–199.

Verspieren M-R. 1990. *Recherche-Action de Type Stratégique et Sciences de L'éducation*. l'Harmattan, Paris.

Vietor DM, Cralle HT. 1992. Value-laden knowledge and holistic thinking in agricultural research. *Agriculture and Human Values* 9(3):44–57.

Vogel J, Krebs P. 1994. Genèse d'une fédération de paysans, pp 773–776. In: Symposium International: Recherches-Système en Agriculture et Développement Rural, November 21–25. CIRAD, Montpellier, France.

Warner M, Robb C, Mackay A, Brocklesby M. 1996. Linking PRA to policy: The conflict analysis framework, pp 42–47. In: PLA Notes, no. 27. International Institute for Environment and Development, London.

Webler T. 1995. Right discourse in citizen participation: An evaluative yardstick, pp 35–77. In: Renn O, Webler T, Wiedemann P, eds. *Fairness and Competence in Citizen Participation*. Kluwer, Dordrecht, The Netherlands.

Wilkins JV. 1991. The search for a viable alternative to slash and burn agriculture in the lowland plains of Bolivia. *Experimental Agriculture* 27:39–46.

William RD, Lev L, Conway F, Deboodt T, Hathaway R, Todd R, and Smith F. 1994. Improving Oregon natural resources: Collaborative learning, systems approaches and participatory action research, pp 355–359. In: Symposium International Recherches-Système en Agriculture et Développement Rural, November 21–25. CIRAD, Montpellier, France.

Wilson EO. 1992. *The Diversity of Life*. Harvard University Press, Cambridge, MA.

Wimsatt WC. 1980. Reductionist research strategies and their biases in the units of selection controversy, pp 155–201. In: Saarinen E, ed. *Conceptual Issues in Ecology*. Reidel, Dordrecht, The Netherlands.

Winch P. 1958. *The Idea of a Social Science*. Routledge and Kegan, London.

Woodward J. 1958. *Management and Technology*. HMSO (Her Majesty's Stationary Office Publications), London.

World Bank. 1992. Brazil: An analysis of environmental problems in the Amazon (103 pp and annexes). Author, Washington, DC.

World Resource Institute. 1990. *World Resources. People and the Environment*. Oxford University Press, Oxford, UK.

World Resource Institute. 1994. *World Resources 1990–1991*. Oxford University Press, New York.

Wright S, Nelson N. 1995. Participatory research and participant observation: Two incompatible approaches, pp 43–60. In: Wright S, Nelson N, eds. *Power and Participatory Development: Theory and Practice*. Intermediate Technology Publications, London.

Young GL. 1992. Between the atom and the void. Hierarchy in human ecology. *Advances in Human Ecology* 1:119–147.

Acronyms and Abbreviations

CAT	Centro Agroambiental do Tocantins (Marabá)
CEPAGRO	Centro de Promoção da Agricultura de Grupo (Florianopolis/Santa Catarina)
CEPLAC	Cocoa Extension and Marketing Board
CFR	Casa Familiar Rural
CIRAD	Centre de Coopération Internationale en Recherche Agronomique pour le Développement
CIMI	Centro Indigeniste Missionario
CNBB	Conferencia Nacional dos Bispos Brasileiros
CPATU	Centro de Pesquisa Agropecuária do Trópico Humido (Belém)
CPT	Comissão Pastoral da Terra
DAZ	Desenvolvimento da Agricultura Familiar Amazonica (Graduate course of NEAF [European Commission])
EMATER	Empresa de Assistência Técnica e Extensão Rural (State Extension Service)
EMBRAPA	Empresa Brasileira de Pesquisa Agropecuaria (National Agricultural Research)
FAO	Food and Agriculture Organization
FNO	Fundo Constitucional do Norte
FSR	Farming Systems Research
FSR/D	Farming Systems Research and Development
FUNAI	Fundação Nacional do Indio

FUNDASUR	Fundação para o Desenvolvimento do Município de Uruará
GRET	Groupe de Recherches et d'Echanges Technologiques (Paris)
IBAMA.	Instituto Brasileiro do Meio Ambiente
IBDF	Instituto Brasileiro Desenvolvimento Florestal (Brazilian Institute for the Development of Forestry)
IITA	International Institute for Tropical Agriculture
IIED	International Institute for Environment and Development (London)
ILEIA	Institute for Low External Input Agriculture (Wageningen)
IMAZON	Instituto do Meio Ambiente Amazônico (Belém)
INCRA	Instituto Nacional da Colonização e Reforma Agraria
INPA	Instituto Nacional de Pesquisa Amazônica (Manaus)
IPCC	Intergovernmental Panel on Climate Changes
ITERPA	Instituto de Terras do Pará
LAET	Laboratorio Agro Ecologico da Transmazônica (Altamira)
MAB	Man and Biosphere; Program of the United Nations Educational, Scientific and Cultural Organization
MPST	Movimento Pela Sobrevivencia da Transamazônica (Altamira)
NAEA	Nucleo de Altos Estudos Amazônicos (Belém)
NEAF	Nucleo de Estudo da Agricultura Familiar Amazônica (Belém)
NGO	nongovernment organization
NRM	natural resource management
ODI	Overseas Development Institute (London)
PAET	Programa Agro-Ecologico da Transamazônica
PAR	Participatory Action Research
PPG7	Pilot Program for Amazonia of the G7 (group of seven developed countries)
PT	Partido dos Trabalhadores
SACTES	Serviço Alemão de Cooperação Tecnica e Social
STR	Sindicato de Trabalhadores Rurais
SUDAM	Superintendencia de Desenvolvimenta da Amazonia (Belém)
UFPa	Universidade Federal do Pará. (Belém)
WWF	World Wide Fund for Nature

2

APPENDIX

LAET Publications

Atas do seminário sobre o uso dos recursos naturais em Porto de Moz (madeira e pesca). 5 a 9/05/96. LAET/MPST/STR/ASPAR/Paróquia de Porto De Moz e CPT (56 pp). LAET, Altamira.

Atas do seminário sobre exploração madeireira e colonização no município de Uruará. Abril/95, 39 pp. Fundasur/LAET, Altamira.

Castellanet C. 1999. L'utilisation de la méthode de "plate-forme de négociation" entre les différents usagers des ressources naturelles dans le cadre de la planification municipale participative. Le cas du travail du LAET avec les municipalités du front pionnier amazonien (Pará-Brésil), pp 536–558. In: Actes du Symposium Jardin planétaire, 14–18 Mars 1999, Chambéry, France.

Castellanet C. 1998. The use of participatory action research for environmental problem-solving. PhD thesis. University of Georgia Institute of Ecology, Athens, Georgia.

Castellanet C, Alves J, David B. 1996. A parceria entre organizações de produtores e equipe de pesquisadores. *Agricultura Familiar: Pesquisa, Formação e Desenvolvimento* 1(1):139–161.

Castellanet C, Alves J, David B, Celestino Filho P, Salgado I, Simões A. 1997. Une nouvelle gestion des ressources naturelles. Le Programme Agro-Ecologique de la Transamazonienne, pp 124–137. In: Thery H, ed. *Environnement et Développement en Amazonie Brésilienne.* Belin, Paris.

Castellanet C, Salgado I, Alves J, Celestino Filho P, Simões A. 1998. La contribution de la recherche-action participative à l'émergence d'un projet collectif de development durable sur la frontière amazonienne, pp 281–300. In: Rossi G, Lavigne Delville P, Narbeburu D, eds. *Sociétés Rurales et Environnement. Gestion des Ressources et Dynamiques Locales au Sud.* Karthala/Regards/GRET, Paris.

Castellanet C, Simões A, Celestino Filho P. 1998. Diagnostico Preliminar da Agricultura Familiar na Transamazônica: Indicações para Pesquisa e Desenvolvimento. Documentos no 105 (48 pp). EMBRAPA/CPATU, Belém (Pará) Brasil.

Clouet Y, Sautier D, Paralieu N. 1996. Fronts pionniers et organisation de l'espace en Amazonie orientale, pp 129–150. In: Albaladejo C, Tulet JC, eds. *Les Fronts Pionniers de l'Amazonie Brésilienne.* L'Harmattan, Paris.

Hebette J. 1996. Relações pesquisadores agricultores. Uma analise estrutural. *Agricultura Familiar: Pesquisa, Formação e Desenvolvimento* 1(1):39–57.

Hebette J, Alves J, Da Silva R. 1996. Parenté, voisinage et organisation professionelle dans la formation du front pionnier amazonien, pp 279–302. In: Albaladejo C, Tulet JC, eds. *Les Fronts Pionniers de l'Amazonie Brésilienne*. L'Harmattan, Paris.

Laigneau M. 1998. Diagnóstico de situação: As lavouras de café, pimenta e cacau na zona leste da Transamazônica, Anapu-Pará. (46 pp). M Sc dissertation Centre National d'Etude de l'Agronomie des Régions Chaudes, Montpellier.

Melo R, Rocha C, Santos MC. 2001. Um aporte metodológico à pesquisa-ação como mecanismo potencializador da regulação do uso dos recursos florestais. O caso das comunidades ribeirinhas do Baixo Rio Xingu. *Agricultura Familiar: Pesquisa, Formação e Desenvolvimento*.

Moreira E, Hebette J, Leitão W. Comunidades ribeirinhas de Porto de Moz e gestão dos recursos naturais. *Agricultura Familiar: Pesquisa, Formação e Desenvolvimento* (in press).

Peixoto LA. Crédito rural para a Agricultura Familiar: O caso do FNO Especial-Prorural na Transamazônica. MSc Dissertation. Núcleo de Estudos da Agricultura Familiar Universidade Federal do Para, Belém (Pará) Brasil.

Rocha, C, Kato A, Flohic A, Reis S, Celestino Filho P, Hoffmann A, Lima H. 2001. Pesquisa ação para revitalização do cultivo da pimenta do reino na Transamazônica. Documentos EMBRAPA/CPATU. Universidade Federal do Pará, Balem, Pará, Brasil.

Salgado I. 1997. L'Exploitation et la Conservation de *Cedrela odorata, Carapa guianensis* et *Swietenia macrophylla* (Meliaceae) en Amazonie Brésilienne. Thèse de Doctorat. Université Paris VI, Paris.

Salgado I, Castellanet C. Contribuição do estudo sobre a exploração madeireira no município de Uruará para uma melhor gestão coletiva dos recursos naturais. *Revista EMBRAPA* (in press).

Salgado I, Castellanet C. 2000. Potencial e limites da pesquisa participativa para o planejamento do uso dos recursos naturais: O caso do município de Uruará: *Agricultura Familiar: Pesquisa, Formação e Desenvolvimento* 1(2):89–111.

Santos MC, Salgado I. 1998. *L'aménagement forestier: Potentialités et limites de son applicabilité en Amazonie*. Communications, I: Les forêts; Dynamiques Sociales et Environnement: Pour un dialogue entre chercheurs, opérateurs et bailleurs de fonds. *Université Bordeaux*, 1:159–170.

Schmitz H. 1996. Desenvolvimento participativo de tecnologias: A experiência da mecanização na Transamazônica. *Agricultura Familiar: Pesquisa, Formação e Desenvolvimento* 1(1):1–20.

Schmitz H, Castellanet C, Simões A. 1996. A participação dos agricultores e das suas organizações no processo de desenvolvimento das tecnologias na região da Transamazônica. *Boletim do Museu Goeldi: Antropologia* 12(2):201–246.

Schmitz H, Simões A, Castellanet C. 1997. Why do farmers experiment with animal traction in Amazonia? In: *Farmers Research in Practice*. Institute for Low External Input Agriculture (Wageningen) /Intermediate Technology Publications, London.

Simões A. 1999. Agricultores e pesquisadores no processo de construção social da demanda de pesquisa-ação. MSc Dissertation. Núcleo de Estudos da Agricultura Familiar Universidade Federal do Pará. Belém (Pará) Brazil.

Simões A. 1996. A construção da pesquisa-desenvolvimento participativa: Reflexões sobre a introdução da mecanização na Transamazônica. *Agricultura Familiar Familiar: Pesquisa, Formação e Desenvolvimento* 1(1).

AUTHOR INDEX

SUBJECT INDEX